D1498128

Writing the Critical Essay

Climate Change

An OPPOSING VIEWPOINTS® Guide

Lauri S. Friedman, *Book Editor*

PROPERTY OF
MONTCALM
COMMUNITY COLLEGE

**OPPOSING
VIEWPOINTS®
SERIES**

GREENHAVEN PRESS
A part of Gale, Cengage Learning

GALE
CENGAGE Learning™

Detroit • New York • San Francisco • New Haven, Conn • Waterville, Maine • London

Christine Nasso, *Publisher*
Elizabeth Des Chenes, *Managing Editor*

© 2009 Greenhaven Press, a part of Gale, Cengage Learning

Gale and Greenhaven Press are registered trademarks used herein under license.

For more information, contact:
Greenhaven Press
27500 Drake Rd.
Farmington Hills, MI 48331-3535
Or you can visit our Internet site at gale.cengage.com

ALL RIGHTS RESERVED.
No part of this work covered by the copyright herein may be reproduced, transmitted, stored, or used in any form or by any means graphic, electronic, or mechanical, including but not limited to photocopying, recording, scanning, digitizing, taping, Web distribution, information networks, or information storage and retrieval systems, except as permitted under Section 107 or 108 of the 1976 United States Copyright Act, without the prior written permission of the publisher.

For product information and technology assistance, contact us at

Gale Customer Support, 1-800-877-4253
For permission to use material from this text or product, submit all requests online at www.cengage.com/permissions

Further permissions questions can be emailed to permissionrequest@cengage.com

Articles in Greenhaven Press anthologies are often edited for length to meet page require-ments. In addition, original titles of these works are changed to clearly present the main thesis and to explicitly indicate the author's opinion. Every effort is made to ensure that Greenhaven Press accurately reflects the original intent of the authors. Every effort has been made to trace the owners of copyrighted material.

Cover image © 2009/Jupiterimages.

LIBRARY OF CONGRESS CATALOGING-IN-PUBLICATION DATA

Climate change / Lauri S. Friedman, book editor.
    p. cm. — (Writing the critical essay, an opposing viewpoints guide)
  Includes bibliographical references and index.
  ISBN 978-0-7377-4402-6 (hardcover)
  1. Climatic changes. 2. Essays—Authorship. I. Friedman, Lauri S.
  QC981.8.C5C21967 2009
  363.738'74—dc22

32.00   2008053350

Printed in the United States of America
1 2 3 4 5 6 7 13 12 11 10 09

Examining the state of writing and how it is taught in the United States was the official purpose of the National Commission on Writing in America's Schools and Colleges. The commission, made up of teachers, school administrators, business leaders, and college and university presidents, released its first report in 2003. "Despite the best efforts of many educators," commissioners argued, "writing has not received the full attention it deserves." Among the findings of the commission was that most fourth-grade students spent less than three hours a week writing, that three-quarters of high school seniors never receive a writing assignment in their history or social studies classes, and that more than 50 percent of first-year students in college have problems writing error-free papers. The commission called for a "cultural sea change" that would increase the emphasis on writing for both elementary and secondary schools. These conclusions have made some educators realize that writing must be emphasized in the curriculum. As colleges are demanding an ever-higher level of writing proficiency from incoming students, schools must respond by making students more competent writers. In response to these concerns, the SAT, an influential standardized test used for college admissions, required an essay for the first time in 2005.

Books in the Writing the Critical Essay: An Opposing Viewpoints Guide series use the patented Opposing Viewpoints format to help students learn to organize ideas and arguments and to write essays using common critical writing techniques. Each book in the series focuses on a particular type of essay writing—including expository, persuasive, descriptive, and narrative—that students learn while being taught both the five-paragraph essay as well as longer pieces of writing that have an opinionated focus. These guides include everything necessary to help students research, outline, draft, edit, and ultimately write successful essays across the curriculum, including essays for the SAT.

Using Opposing Viewpoints

This series is inspired by and builds upon Greenhaven Press's acclaimed Opposing Viewpoints series. As in the parent series, each book in the Writing the Critical Essay series focuses on a timely and controversial social issue that provides lots of opportunities for creating thought-provoking essays. The first section of each volume begins with a brief introductory essay that provides context for the opposing viewpoints that follow. These articles are chosen for their accessibility and clearly stated views. The thesis of each article is made explicit in the article's title and is accentuated by its pairing with an opposing or alternative view. These essays are both models of persuasive writing techniques and valuable research material that students can mine to write their own informed essays. Guided reading and discussion questions help lead students to key ideas and writing techniques presented in the selections.

The second section of each book begins with a preface discussing the format of the essays and examining characteristics of the featured essay type. Model five-paragraph and longer essays then demonstrate that essay type. The essays are annotated so that key writing elements and techniques are pointed out to the student. Sequential, step-by-step exercises help students construct and refine thesis statements; organize material into outlines; analyze and try out writing techniques; write transitions, introductions, and conclusions; and incorporate quotations and other researched material. Ultimately, students construct their own compositions using the designated essay type.

The third section of each volume provides additional research material and writing prompts to help the student. Additional facts about the topic of the book serve as a convenient source of supporting material for essays. Other features help students go beyond the book for their research. Like other Greenhaven Press books, each book in the Writing the Critical Essay series includes bibliographic listings of relevant periodical articles, books, Web sites, and organizations to contact.

Writing the Critical Essay: An Opposing Viewpoints Guide will help students master essay techniques that can be used in any discipline.

The Global Nature of Global Warming

A most impressive—and daunting—facet of climate change is its potential to fundamentally affect every corner of the planet. Indeed, the implications of global warming are truly global, though not necessarily in equal ways. While all nations are sure to be affected by global warming should it occur on the scale and at the rate that experts predict, the quality of that change will fluctuate in different areas of the world. Some areas will suffer more than others, and some areas might even benefit from climate change.

A country might benefit from climate change if a warmer climate exposes riches currently locked up in cold and ice. For example, countries that possess large swaths of land that are currently uninhabitable might see a temperature increase peel back layers of snow and ice to reveal land that is fertile and livable. Greenland is one place where retreating glaciers have already extended the growing season by two weeks a year and opened previously frozen areas for exploration and tourism. Russia's Siberia is another such place: Should global warming melt the vast frozen tundra there, it could reveal untapped oil deposits, fertile soil for agriculture, and thousands and thousands of miles of land that could be profitably developed. "Climate change [could] increase the supply of land by warming currently frosty areas," writes Brookings Institution fellow Gregg Easterbrook. "Canada, Greenland, and Scandinavia [could] experience a rip-roarin' economic boom [while] warming's benefits to Russia could exceed those to all other nations combined."[1]

While the fight against global warming is typically perceived as a global endeavor, one in which all countries of the world have a stake, it is interesting to consider that from the perspective of nations that stand to benefit from climate change, it might be against their national interest to halt it. Indeed, Easterbrook suggests this was the thinking behind Russia's reluctance to sign the Kyoto Protocol, the international document whose aim is to limit the global emission of climate-changing greenhouse gases. It might also be the reason behind Canada's recent increase in greenhouse gas emissions: "Maybe this is a result of prosperity and oil-field development," speculates Easterbrook, "or maybe those wily Canadians have a master plan for their huge expanse of currently uninhabitable land."[2]

But while it is possible that some nations stand to gain from climate change, others are sure to suffer from it. In fact, just a few degrees change in the global average temperature is expected to bring famine, drought, flooding, and desertification to areas on every continent, radically altering those places' abilities to provide food, shelter, and clean drinking water for their residents. This devastation is expected to occur on such a mammoth scale that even if a few nations benefited briefly from climate change, the global economy would plunge into a chaos that would mitigate any benefits enjoyed. Furthermore, if currently harsh regions were transformed into temperate, comfortable places to live, they would likely be overrun by refugees from no-longer-habitable places.

And while some places on Earth will suffer from climate change, others may be completely wiped out by it. Consider the case of Tuvalu, one of the world's smallest countries. Tuvalu is an island nation in the Pacific between Hawaii and Australia. Its ten thousand residents live on about ten square miles of low-lying land. In 2002 Tuvalu led an effort of several island nations to potentially sue polluters such as the United States and Australia for contributing to global warming, claiming that such pollution directly endangers their existence by threatening to raise sea levels so high their islands would be drowned.

Thus far the ocean level around Tuvalu has not risen abnormally, but climate change models predict that global warming will melt glaciers and ice sheets into the ocean, raising sea levels as much as twenty-three feet (7m). For places like Tuvalu, where the highest point is about sixteen feet (4.9m) above sea level, such a change would be devastating: Tuvalu would be completely submerged. Said Koloa Talake, an elder statesman on the island, "We don't have hills or mountains. All we have is coconut trees. If the industrial countries don't consider our crisis, our only alternative is to climb up in the coconut trees when the tide rises."[3] That Tuvalu would suffer so much as a result of climate change while it, as a nonindustrial country, has done relatively nothing to contribute to the phenomenon, was the main reasoning behind the idea of the lawsuit.

Other communities around the world are beginning to think about the problem of climate change in the same

A warmer climate change may reveal fertile soil and untapped resources in countries currently covered with snow and ice, such as Greenland.

Scientists point to melting glaciers in New Zealand as evidence of climate change.

way. In 2008, for example, Native Americans living in the Alaskan village of Kivalina filed a lawsuit against corporations such as Exxon, claiming that pollution outputs from the corporations' products (oil and gas) were responsible for environmental damage to their village that threatened their way of life. During fierce winter storms, Kivalina is protected by walls of sea ice that prevent waves from battering its coast. But as a result of climate change, "we are seeing accelerated erosion because of the loss of sea ice," said city administrator Janet Mitchell. "We normally have ice starting in October, but now we have open water even into December so our island is not protected from the storms."[4] The villagers are seeking damages of $400 million, the estimated cost of relocating them to a new piece of land.

Whether the people who live in Kivalina or Tuvalu will be successful in their efforts to hold polluters responsible for climate change–related damage to their land remains to be seen. But it is clear that such lawsuits have heralded a new chapter in the climate change debates. Thinking about climate change as an issue that can be litigated—such as tobacco use or fast-food consumption—is a distinctly new way of trying to curb it and assign responsibility for it.

Taking responsibility for climate change is just one of the issues explored in *Writing the Critical Essay: Climate Change*. Readers will consider arguments on the reality, causes, and effects of climate change and learn to form their own opinions on the matter. Carefully annotated model essays and thought-provoking writing exercises help readers write their own compare-and-contrast essays on this complicated and increasingly important topic.

Notes

1. Gregg Easterbrook, "Global Warming: Who Loses—and Who Wins?" *Atlantic Monthly*, April 2007, pp. 53–54.

2. Easterbrook, "Global Warming: Who Loses—and Who Wins?" p. 54.

3. Quoted in Richard C. Paddock, "Tuvalu's Sinking Feeling," *Los Angeles Times*, October 4, 2002, p. A1.

4. Quoted in Associated Press, "Alaska Town Sues Oil and Power Companies over Global Warming," February 27, 2008.

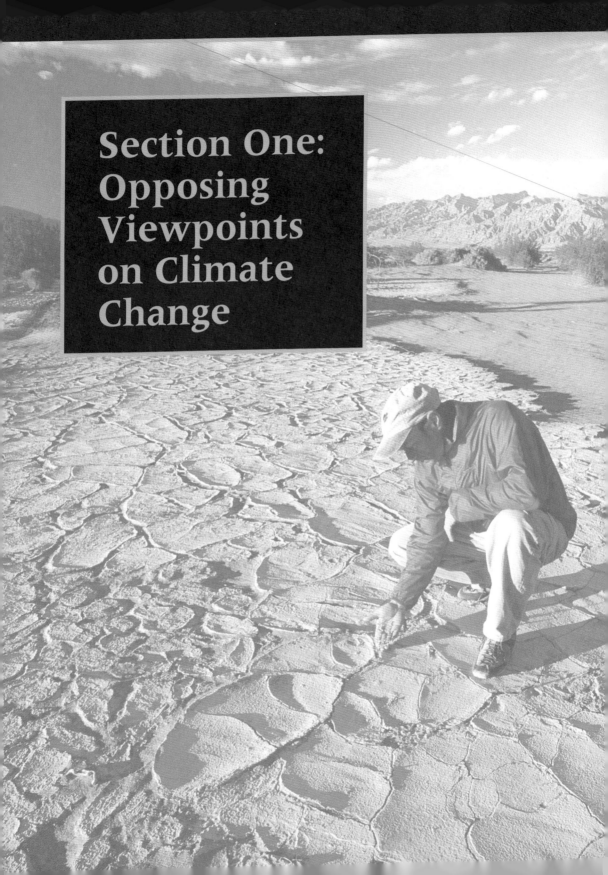

Section One:
Opposing
Viewpoints
on Climate
Change

Climate Change Is a Serious Problem

Peter Backlund, Anthony Janetos, and David Schimel

In the following essay scientists Peter Backlund, Anthony Janetos, and David Schimel argue that climate change is already causing serious problems for the planet's delicate ecosystems. They explain that recent global warming—which is occurring faster than at any point in the last ten thousand years—will have significant agricultural consequences, making it more difficult to sustain commercial crops. Weeds and pests will flourish in a warmer climate, making crops and animals more susceptible to disease and invasive species. Fires will decimate open spaces that are not sufficiently covered with vegetation, and warming temperatures will melt snowpacks, leading to increased runoff and flooding. Species are likely to become extinct, and the authors point out that many coral populations have already demonstrated they are under severe stress from a warmer ocean. Backlund, Janetos, and Schimel conclude that climate change is a serious problem that needs immediate international attention.

Backlund is with the National Center for Atmospheric Research (NCAR), Janetos is a researcher at the University of Maryland, and Schimel works at the National Ecological Observatory Network.

Consider the following questions:

1. By how much will the average global temperature rise by 2100, according to the authors?
2. In what way will a warmer world allow pests and weeds to flourish, according to the authors?
3. What effect do the authors say climate change will have on the frequency of fires?

Peter Backlund, Anthony Janetos, and David Schimel, "The Effects of Climate Change on Agriculture, Land Resources, Water Resources, and Biodiversity: Executive Summary," in usda.gov, U.S. Department of Agriculture, May 2008.

There is a robust scientific consensus that human-induced climate change is occurring. The Fourth Assessment Report (AR4) of the IPCC [Intergovernmental Panel on Climate Change], the most comprehensive and up-to-date scientific assessment of this issue, states with "very high confidence" that human activities, such as fossil fuel burning and deforestation, have altered the global climate.

Our Climate Is Changing Too Rapidly

During the 20th century, the global average surface temperature increased by about 0.6°C and global sea level increased by about 15 to 20 cm. Global precipitation over land increased about two percent during this same period. Looking ahead, human influences will continue to change Earth's climate throughout the 21st century. The IPCC AR4 projects that the global average temperature will rise another 1.1 to 5.4°C by 2100, depending on how much the atmospheric concentrations of greenhouse gases increase during this time. This temperature rise will result in continued increases in sea level and overall rainfall, changes in rainfall patterns and timing, and decline in snow cover, land ice, and sea ice extent. It is very likely that the earth will experience a faster rate of climate change in the 21st century than seen in the last 10,000 years. . . .

Crops Will Fail

Weather and climate factors such as temperature, precipitation, CO_2 concentrations, and water availability directly impact the health and well-being of plants, pasture, rangeland, and livestock. For any agricultural commodity, variation in yield between years is related to growing-season weather; weather also influences insects, disease, and weeds, which in turn affect agricultural production.

With increased CO_2, and temperature, the life cycle of grain and oilseed crops will likely progress more rapidly.

But, as temperature rises, these crops will increasingly begin to experience failure, especially if climate variability increases and precipitation lessens or becomes more variable.

The marketable yield of many horticultural crops— e.g., tomatoes, onions, fruits—is very likely to be more sensitive to climate change than grain and oilseed crops.

Scientists say that climate change is causing more forest fires.

The Consequences of Climate Change

Table 3.2. Examples of possible impacts of climate change due to changes late 21st century. These do not take into account any changes or develop-the phenomena listed in column one. {WGII Table SPM. 1}

Phenomenon[a] and direction of trend	LIkelihood of future trends based on projections for 21st century using SRES scenarios	Examples of major Agriculture, forestry and ecosystems {WGII 4.4, 5.4}
Over most land areas, warmer and fewer cold days and nights, warmer and more frequent hot days and nights	*Virtually certain[b]*	Increased yields in colder environments; decreased yields in warmer environments; increased insect outbreaks
Warm spells/heat waves. Frequency increases over most land areas	*Very likely*	Reduced crop yields in warmer regions due to heat stress; increased danger of wildfire
Heavy precipitation events. Frequency increases over most areas	*Very likely*	Damage to crops; soil erosion, inability to cultivate land due to waterlogging of soils
Area affected by drought increases	*Likely*	Land degradation; lower yields/crop damage and failure; increased livestock deaths; increased risk of wildfire
Intense tropical cyclone activity increases	*Likely*	Damage to crops; windthrow (uprooting) of trees; damage to coral reefs
Increased incidence of extreme high sea level (excludes tsunamis)[c]	*Likely[d]*	Salinisation of irrigation water, estuaries and fresh-water systems

Notes:
a) See WGI Table 3.7 for further details regarding definitions.
b) Warming of the most extreme days and nights each year.
c) Extreme high sea level depends on average sea level and on regional weather systems. It is best defined as the highest 1% of hourly values of observed sea level at a station for a given reference period.
d) In all scenarios, the projected global average sea level at 2100 is higher than in the reference period. The effect of changes in regional weather systems or sea level extremes has not been assessed. {WBI 10.6}

in extreme weather and climate events, based on projections to the mid- to
ments in adaptive capacity. The likelihood estimates in column two relate to

projected impacts by sector

Water resources {WGII 3.4}	Human health {WBII 3.4}	Industry, settlement and society {WBII 7.4}
Effects on water resources relying on snowmelt; effects on some water supplies	Reduced human mortality from decreased cold exposure	Reduced energy demand for heating; increased demand for cooling; declining air quality in cities; reduced disruption to transport due to snow, ice; effects on winter tourism
Increased water demand; water quality problems, e.g. algal blooms	Increased risk of heat-related mortality, especially for the elderly, chronically sick, very young and socially isolated	Reduction in quality of life for people in warm areas without appropriate housing; impacts on the elderly, very young and poor
Adverse effects on quality of surface and groundwater; contamination of water supply; water scarcity may be relieved	Increased risk of deaths, injuries and infectious, respiratory and skin diseases	Disruption of settlements, commerce, transport and societies due to flooding: pressures on urban and rural infrastructures; loss of property
More widespread water stress	Increased risk of food and water shortage; increased risk of malnutrition; increased risk of water- and food-borne diseases	Water shortage for settlements, industry and societies; reduced hydropower generation potentials; potential for population migration
Power outages causing disruption of public water supply	Increased risk of deaths, injuries, water- and food-borne diseases; post-traumatic stress disorders	Disruption by flood and high winds; withdrawal of risk coverage in vulnerable areas by private insurers; potential for population migrations; loss of property
Decreased fresh-water availability due to saltwater intrusion	Increased risk of deaths and injuries by drowning in floods; migration-related health effects	Costs of coastal protection versus costs of land-use relocation; potential for movement of populations and infrastructure; also see tropical cyclones above

Taken from: Lenny Bernstein, et al, *Climate Change and Its Impacts in the Near and Long Term Under Different Scenario*. Geneva 2, Switzerland: Intergovernmental Panel on Climate Change, 2005. Copyright © 2005, Intergovernmental Panel on Climate Change. Reproduced by permission.

Weeds and Pests Will Flourish

Climate change is likely to lead to a northern migration of weeds. Many weeds respond more positively to increasing CO_2 than most cash crops, particularly C3 "invasive" weeds. Recent research also suggests that glyphosate, the most widely used herbicide in the United States, loses its efficacy on weeds grown at the increased CO_2 levels likely in the coming decades.

Disease pressure on crops and domestic animals will likely increase with earlier springs and warmer winters, which will allow proliferation and higher survival rates of pathogens and parasites. Regional variation in warming and changes in rainfall will also affect spatial and temporal distribution of disease.

Projected increases in temperature and a lengthening of the growing season will likely extend forage production into late fall and early spring, thereby decreasing need for winter season forage reserves. However, these benefits will very likely be affected by regional variations in water availability.

Climate change–induced shifts in plant species are already under way in rangelands. Establishment of perennial herbaceous species is reducing soil water availability early in the growing season. Shifts in plant productivity and type will likely also have significant impact on livestock operations. . . .

Changes in Forests and Arid Lands

Climate strongly influences forest productivity, species composition, and the frequency and magnitude of disturbances that impact forests. The effect of climate change on disturbances such as forest fire, insect outbreaks, storms, and severe drought will command public attention and place increasing demands on management resources. Disturbance and land use will control the response of arid lands to climate change. Many plants and animals in arid ecosystems are near their physiological limits for tolerating temperature and water stress and even slight changes in stress will have significant consequences. In the near term, fire effects will trump climate effects on ecosystem structure and function.

More Fires, Thinner Forests

Climate change has very likely increased the size and number of forest fires, insect outbreaks, and tree mortality in the interior West, the Southwest, and Alaska, and will continue to do so.

Rising CO_2 will very likely increase photosynthesis for forests, but this increase will likely only enhance wood production in young forests on fertile soils. . . .

Higher temperatures, increased drought, and more intense thunderstorms will very likely increase erosion and promote invasion of exotic grass species in arid lands.

Climate change in arid lands will create physical conditions conducive to wildfire, and the proliferation of exotic grasses will provide fuel, thus causing fire frequencies to increase in a self-reinforcing fashion.

In arid regions where ecosystems have not coevolved with a fire cycle, the probability of loss of iconic, charismatic megaflora such as saguaro cacti and Joshua trees is very likely.

Arid lands very likely do not have a large capacity to absorb CO_2 from the atmosphere and will likely lose carbon as climate-induced disturbance increases.

> ### The Climate Has Already Started to Change
>
> Nearly every research institution involved in the study of global climate change—from the American Academy of Sciences to the atmospheric department at your local university—has issued reports citing overwhelming evidence that the planet is changing. . . . Don't think of it in terms of centuries, or even decades. It's happening now.
>
> Lee Dye, "Global Climate Change Is Happening Now: Scientists Fear Global Warming Is Irreversible and Its Effects Possibly Disastrous," ABC News, July 12, 2006. http://abcnews.go.com/Technology/Story?id = 2182824&page = 1.

Vegetation Loss Is Likely

River and riparian ecosystems in arid lands will very likely be negatively impacted by decreased streamflow, increased water removal, and greater competition from non-native species.

An 89°F thermometer reading on top of mountain snow fields highlights the trend toward reduced mountain snow packs and early spring snowmelt runoffs in the western United States.

Changes in temperature and precipitation will very likely decrease the cover of vegetation that protects the ground surface from wind and water erosion.

Current observing systems do not easily lend themselves to monitoring change associated with disturbance and alteration of land cover and land use, and distinguishing such changes from those driven by climate change. Adequately distinguishing climate change influences is aided by the collection of data at certain spatial

and temporal resolutions, as well as supporting ground truth measurements. . . .

Climate Change Will Affect Our Water Supply

Plants, animals, natural and managed ecosystems, and human settlements are susceptible to variations in the storage, fluxes, and quality of water, all of which are sensitive to climate change. The effects of climate on the nation's water storage capabilities and hydrologic functions will have significant implications for water management and planning as variability in natural processes increases. Although U.S. water management practices are generally quite advanced, particularly in the West, the reliance on past conditions as the foundation for current and future planning and practice will no longer be tenable as climate change and variability increasingly create conditions well outside of historical parameters and erode predictability.

Most of the United States experienced increases in precipitation and streamflow and decreases in drought during the second half of the 20th century. It is likely that these trends are due to a combination of decadal-scale variability and long-term change.

Consistent with streamflow and precipitation observations, most of the continental United States experienced reductions in drought severity and duration over the 20th century. However, there is some indication of increased drought severity and duration in the western and southwestern United States.

Increase Melting and Runoff

There is a trend toward reduced mountain snowpack and earlier spring snowmelt runoff peaks across much of the western United States. This trend is very likely attributable at least in part to long-term warming, although some part may have been played by decadal-scale variability, including a shift in the phase of the Pacific Decadal Oscillation in the late 1970s. Where earlier snowmelt peaks and reduced summer and fall low flows have already been detected, continuing shifts in this direction are very likely and may have substantial impacts on the performance of reservoir systems. . . .

Species Will Be Affected and Threatened

Biodiversity, the variation of life at the genetic, species, and ecosystem levels of biological organization, is the fundamental building block of the services that ecosystems deliver to human societies. It is intrinsically important both because of its contribution to the functioning of ecosystems, and because it is difficult or impossible to recover or replace, once it is eroded. Climate change is affecting U.S. biodiversity and ecosystems, including changes in growing season, phenology, primary production, and species distributions and diversity. It is very likely that climate change will increase in importance as a driver for changes in biodiversity over the next several decades, although for most ecosystems it is not currently the largest driver of change. . . .

Subtropical and tropical corals in shallow waters have already suffered major bleaching events that are clearly driven by increases in sea surface temperatures. Increases in ocean acidity, which are a direct consequence of increases in atmospheric carbon dioxide, are calculated to have the potential for serious negative consequences for corals.

The rapid rates of warming in the Arctic observed in recent decades, and projected for at least the next century, are dramatically reducing the snow and ice covers that provide denning and foraging habitat for polar bears.

There are other possible, and even probable, impacts and changes in biodiversity (e.g., disruption of the relationships between pollinators, such as bees, and flowering plants), for which we do not yet have a substantial observational database. However, we cannot conclude that the lack of complete observations is evidence that changes are not occurring.

Analyze the essay:

1. The authors of this essay, Peter Backlund, Anthony Janetos, and David Schimel, are scientists who work at a variety of academic and government agencies. Does knowing the background of these authors influence your opinion of their arguments? If so, in what way?

2. Backlund, Janetos, and Schimel assert that climate change is occurring and warn it will have serious repercussions for humans and the environment. How do you think each of the authors in this section would respond to this claim? After you read each essay, write two or three sentences on what each author would say about Backlund, Janetos, and Schimel's claims.

Climate Change Is Not a Serious Problem

Christopher Monckton

In the following essay Christopher Monckton argues that climate change is a myth perpetrated by hysterical government officials. He challenges the assertions of the United Nations and other organizations that have argued that the planet is warming with disastrous consequences. Monckton claims these organizations have used faulty data to make their conclusions. In some cases they have not measured correctly; in other cases they have consciously chosen to ignore data that contradict their conclusions. Monckton contends that the earth has endured warmer periods in the past and that today's warming is likely attributable to the sun, not CO_2 emissions from human activity. Furthermore, he claims that many areas on the planet are either cooling or remain unchanged in temperature. For all these reasons, Monckton urges people to not believe hype and hysteria about climate change.

Monckton is a British politician and business consultant, policy adviser, writer, and inventor. He is well known for his public opposition to the mainstream scientific consensus on global warming and climate change.

Consider the following questions:
1. What is wrong with the UN's "hockey stick" graph, according to Monckton?
2. How many degrees warmer than current temperatures does the author say the medieval warm period was?
3. What have the last thirty years shown about the Antarctic, according to Monckton?

Christopher Monckton, "Climate Chaos? Don't Believe It," Telegraph.co.uk, May 11, 2006. Reproduced by permission.

[I]n 2006], Gordon Brown [prime minister of the United Kingdom] and his chief economist both said global warming was the worst "market failure" ever. That loaded soundbite suggests that the "climate-change" scare is less about saving the planet than, in [former French president] Jacques Chirac's chilling phrase, "creating world government". . . . I'll reveal how politicians, scientists and bureaucrats contrived a threat of Biblical floods, droughts, plagues, and extinctions worthier of St John the Divine than of science.

Sir Nicholas Stern's report on the economics of climate change, which was published [in May 2006], says that the debate is over. It isn't. There are more greenhouse gases in the air than there were, so the world should warm a bit, but that's as far as the "consensus" goes. After the recent hysteria, you may not find the truth easy to believe. . . .

The Experts Have Tampered with the Data

So to the scare. First, the UN implies that carbon dioxide ended the last four ice ages. It displays two 450,000-year graphs: a sawtooth curve of temperature and a sawtooth of airborne CO_2 that's scaled to look similar. Usually, similar curves are superimposed for comparison. The UN didn't do that. If it had, the truth would have shown: the changes in temperature preceded the changes in CO_2 levels.

Next, the UN abolished the medieval warm period (the global warming at the end of the First Millennium AD). In 1995, David Deming, a geoscientist at the University of Oklahoma, had written an article reconstructing 150 years of North American temperatures from borehole data. He later wrote: "With the publication of the article in *Science*, I gained significant credibility in the community of scientists working on climate change. They thought I was one of them, someone who would pervert science in the service of social and political causes. One of them let his guard down. A major person working in the area of climate change and global warming sent me an astonishing email that said: 'We have to get rid of the Medieval Warm Period.'"

An Incorrect Model

So they did. The UN's second assessment report, in 1996, showed a 1,000-year graph demonstrating that temperature in the Middle Ages was warmer than today. But the 2001 report contained a new graph showing no medieval warm period. It wrongly concluded that the 20th century was the warmest for 1,000 years. The graph looked like an ice hockey–stick. The wrongly flat AD1000–AD1900 temperature line was the shaft: the uptick from 1900 to 2000 was the blade. Here's how they did it:

- They gave one technique for reconstructing pre-thermometer temperature 390 times more weight than any other (but didn't say so).
- The technique they overweighted was one which the UN's 1996 report had said was unsafe: measurement of tree-rings from bristlecone pines. Tree-rings are wider in warmer years, but pine-rings are also wider when there's more carbon dioxide in the air: it's plant food. This carbon dioxide fertilisation distorts the calculations.
- They said they had included 24 data sets going back to 1400. Without saying so, they left out the set showing the medieval warm period, tucking it into a folder marked "Censored Data".
- They used a computer model to draw the graph from the data, but scientists later found that the model almost always drew hockey-sticks even if they fed in random, electronic "red noise".

The World Has Been Warmer in the Past

The large, full-colour "hockey-stick" was the key graph in the UN's 2001 report, and the only one to appear six times. The Canadian Government copied it to every household. Four years passed before a leading scientific

journal would publish the truth about the graph. Did the UN or the Canadian government apologise? Of course not. The UN still uses the graph in its publications.

Even after the "hockey stick" graph was exposed, scientific papers apparently confirming its abolition of the medieval warm period appeared. The US Senate asked independent statisticians to investigate. They found that the graph was meretricious, and that known associates of the scientists who had compiled it had written many of the papers supporting its conclusion.

The UN, echoed by Stern, says the graph isn't important. It is. Scores of scientific papers show that the medieval warm period was real, global and up to 3C warmer than now. Then, there were no glaciers in the tropical Andes: today they're there. There were Viking farms in Greenland: now they're under permafrost. There was little ice at the North Pole: a Chinese naval squadron sailed right round the Arctic in 1421 and found none.

The United Nations' 1996 report on global warming ignores data from tree rings indicating carbon dioxide fertilization, which distorts their calculations.

Some Areas of the World Are Cooling

The Antarctic, which holds 90 per cent of the world's ice and nearly all its 160,000 glaciers, has cooled and

Some believe that vanishing snow packs on Mount Kilimanjaro in Tanzania, Africa, are caused by dry air from deforestation, not climate change.

gained ice-mass in the past 30 years, reversing a 6,000-year melting trend. Data from 6,000 boreholes worldwide show global temperatures were higher in the Middle Ages than now. And the snows of Kilimanjaro are vanishing not because summit temperature is rising (it isn't) but because post-colonial deforestation has dried the air. . . .

The Sun, Not CO₂, Could Be Causing Warming

In some places it was also warmer than now in the Bronze Age and in Roman times. It wasn't CO_2 that caused those warm periods. It was the sun. So the UN adjusted the maths and all but extinguished the sun's role in today's warming. Here's how:

- The UN dated its list of "forcings" (influences on temperature) from 1750, when the sun, and consequently air temperature, was almost as warm as now. But its start-date for the increase in world temperature was 1900, when the sun, and temperature, were much cooler.
- Every "forcing" produces "climate feedbacks" making temperature rise faster. For instance, as temperature rises in response to a forcing, the air carries

more water vapour, the most important greenhouse gas; and polar ice melts, increasing heat absorption. Up goes the temperature again. The UN more than doubled the base forcings from greenhouse gases to allow for climate feedbacks. It didn't do the same for the base solar forcing. . . .

The UN expresses its heat energy forcings in watts per square metre per second. It estimates that the sun caused just 0.3 watts of forcing since 1750. Begin in 1900 to match the temperature start-date, and the base solar forcing more than doubles to 0.7 watts. Multiply by 2.7, which the Royal Society suggests is the UN's current factor for climate feedbacks, and you get 1.9 watts—more than six times the UN's figure.

The entire 20th-century warming from all sources was below 2 watts. The sun could have caused just about all of it.

Guesswork Rather than Facts

Next, the UN slashed the natural greenhouse effect by 40 per cent from 33C in the climate-physics textbooks to 20C, making the man-made additions appear bigger.

Then the UN chose the biggest 20th-century temperature increase it could find. Stern says: "As anticipated by scientists, global mean surface temperatures have risen over the past century." As anticipated? Only 30 years ago, scientists were anticipating a new Ice Age and writing books called *The Cooling*.

In the US, where weather records have been more reliable than elsewhere, 20th-century temperature went up by only 0.3C. AccuWeather, a worldwide meteorological service, reckons world temperature rose by 0.45C. The US National Climate Data Centre says 0.5C. Any advance

> ### Global Warming Could Help Crops Grow
> Global warming increases carbon dioxide (CO_2), which acts like fertilizer for plants. . . . Since 1950, in a period of global warming, these factors have helped the world's grain production soar from 700 million to more than 2 billion tons.
>
> Dennis T. Avery and H. Sterling Burnett, "Global Warming: Famine—or Feast?" National Center for Policy Analysis, Brief Analysis no. 517, May 19, 2005. www.ncpa.org/pub/ba/ba517/.

© 2007 by Eric Allie and PoliticalCartoons.com.

on 0.5? The UN went for 0.6C, probably distorted by urban growth near many of the world's fast-disappearing temperature stations.

The number of temperature stations round the world peaked at 6,000 in 1970. It's fallen by two-thirds to 2,000 now: a real "hockey-stick" curve, and an instance of the UN's growing reliance on computer guesswork rather than facts. . . .

Ocean Temperature and Level Are Unchanged

The oceans, we're now told, are acting as a giant heat-sink. In these papers the well-known, central flaw (not mentioned by Stern) is that the computer models' "predictions" of past ocean temperature changes only approach reality if they are averaged over a depth of at least a mile and a quarter.

Deep-ocean temperature hasn't changed at all, it's barely above freezing. The models tend to over-predict the warming of the climate-relevant surface layer up to threefold. A recent paper by John Lyman, of the US National Oceanic and Atmospheric Association, reports that the oceans have cooled sharply in the past two years. The computers didn't predict this. Sea level is scarcely rising faster today than a century ago: an inch every 15 years. [U.S. climatologist James] Hansen now says that the oceanic "flywheel effect" gives us extra time to act, so Stern's alarmism is misplaced.

Finally, the UN's predictions are founded not only on an exaggerated forcing-to-temperature conversion factor justified neither by observation nor by physical law, but also on an excessive rate of increase in airborne carbon dioxide. The true rate is 0.38 per cent year on year since records began in 1958. The models assume 1 per cent per annum, more than two and a half times too high. In 2001, the UN used these and other adjustments to predict a 21st-century temperature increase of 1.5 to 6C. Stern suggests up to 10C. . . .

Why haven't air or sea temperatures turned out as the UN's models predicted? Because the science is bad, the "consensus" is wrong.

Analyze the essay:

1. In this essay Christopher Monckton accuses scientific organizations of using faulty data and exaggerating the severity of climate change. Did he convince you to agree with him? Why or why not? Explain your answer thoroughly.

2. Monckton says that increased CO_2 in the air is like plant food, helping plants grow rather than hurting them. How do you think Peter Backlund, Anthony Janetos, and David Schimel, the authors of the previous essay, would respond to this claim?

An international panel of climate scientists said yesterday that there is an overwhelming probability that human activities are warming the planet at a dangerous rate, with consequences that could soon take decades or centuries to reverse.

"Ninety Percent Certain"

The Intergovernmental Panel on Climate Change [IPCC], made up of hundreds of scientists from 113 countries, said that based on new research over the last six years, it is 90 percent certain that human-generated greenhouse gases account for most of the global rise in temperatures over the past half-century.

Declaring that "warming of the climate system is unequivocal," the authors said in their "Summary for Policymakers" that even in the best-case scenario, temperatures are on track to cross a threshold to an unsustainable level. A rise of more than 3.6 degrees Fahrenheit above pre-industrial levels would cause global effects—such as massive species extinctions and melting of ice sheets—that could be irreversible within a human lifetime. Under the most conservative IPCC scenario, the increase will be 4.5 degrees by 2100.

Severe Global Warming Is at Hand

Richard Somerville, a distinguished professor at the Scripps Institution of Oceanography and one of the lead authors, said the world would have to undertake "a really massive reduction in emissions," on the scale of 70 to 80 percent, to avert severe global warming.

The scientists wrote that it is "very likely" that hot days, heat waves and heavy precipitation will become more frequent in the years to come, and "likely" that future tropical hurricanes and typhoons will become more intense. Arctic sea ice will disappear "almost entirely" by the end of the century, they said, and snow cover will contract worldwide.

Warming Is Human-Caused

While the summary did not produce any groundbreaking observations—it reflects a massive distillation of the peer-reviewed literature through the middle of 2006—it represents the definitive international scientific and political consensus on climate science. It provides much more definitive conclusions than the panel's previous report in 2001, which said only that it was "likely"—meaning between 66 and 90 percent probability on a scale the panel adopted—that human activity accounted for the warming recorded over the past 50 years.

Some of the report's most compelling sections focused on future climate changes, because the buildup of carbon dioxide in the atmosphere would exert an effect even if industrialized countries stopped emitting greenhouse gases tomorrow. Gerald Meehl, a senior scientist at the National Center for Atmospheric Research [NCAR] in Boulder, Colo., who helped oversee the chapter on climate projections, said that in the next two decades alone, global temperatures will rise by 0.7 degrees Fahrenheit.

"We're committed to a certain amount of warming," said Meehl, who worked with 16 computer-modeling teams from 11 countries. "A lot of these changes continue through the 21st century and become more severe as time goes on."

Meehl added, however, that a sharp cut in greenhouse gas emissions could still keep catastrophic consequences from occurring: "The message is, it does make a difference what we do."

For the first time, IPCC scientists also looked at regional climate shifts in detail, concluding that precipitation in the American Southwest will decline as summer temperatures rise, just as precipitation in the Northeast will increase. Linda Mearns—another NCAR senior scientist who was also one of the lead authors—said these changes could cause water shortages and affect recreational activities in the Southwest. Developing countries in Africa and elsewhere could also experience severe droughts.

Warmer Climate, Rising Seas

Governments and scientific organizations across the globe nominate scientists to produce and review the IPCC assessment without pay under the auspices of the United Nations. A group of key authors and government officials met in Paris [in January 2007] to finalize the document, which reflects three years of work.

"Every government in the world signed off on this document, including the U.S.," said World Bank chief scientist Robert T. Watson, who chaired the last round of deliberations. Watson added that compared with the 2001 report, "the difference is now they have more confidence in what they're doing."

The authors concluded that Earth's average temperature will increase between 3.2 and 7.8 degrees Fahrenheit over the next century, while sea levels will rise between seven and 23 inches.

In February 2007 the United Nations Intergovernmental Panel on Climate Change (IPCC) released a report that said there was a 90 percent chance that global warming is man-made.

Man-Made Warming Is Occurring Around the Planet

Models that take into account anthropogenic—or human-caused—warming show a significant temperature increase in all areas of the world.

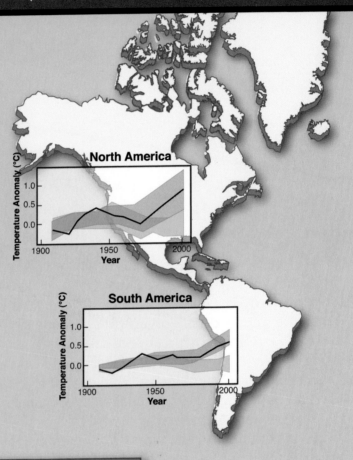

North America

South America

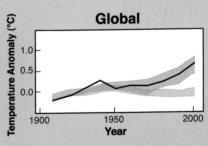

Models using only natural influences on temperature

Models using both natural and anthropogenic influences on temperature

Global

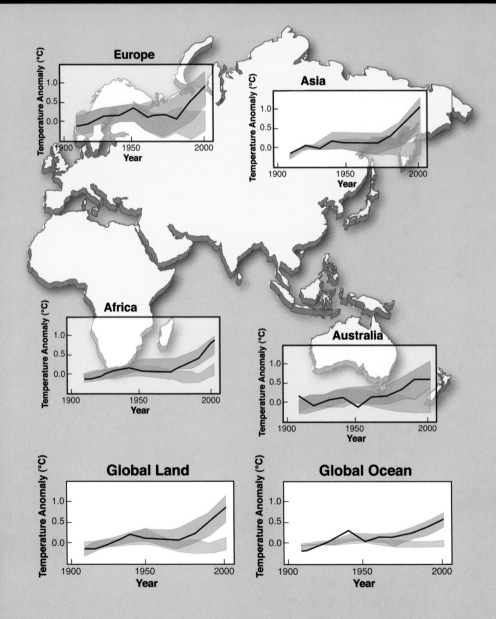

Taken from: Gary W. Yohe, Richard S.J. Tol, Richard G. Richels, and Geoffrey J. Blanford, "Copenhagen Consensus 2008 Challenge Paper: Global Warming," Copenhagen Consensus Center, April 3, 2008.

IPCC scientists also said that global warming will not trigger a shutdown within the next 100 years of the North Atlantic ocean current that keeps Northern Europe temperate, though they do not predict whether it might occur in future centuries. In a similar vein, the authors said they did not have sophisticated enough computer models to project how much melting of the Greenland ice sheet would boost sea levels over the next century, but they suggested that over several centuries the ice sheet's disappearance could raise sea levels by a devastating 23 feet.

Bush administration officials said yesterday that they welcomed the report and emphasized that U.S. research funding helped underpin its conclusions. National Oceanic and Atmospheric Administration Administrator Conrad C. Lautenbacher Jr., who oversees much of the nation's climate research, said in an interview that the international assessment will lead to "a more objective and informative public debate."

The Current Warming Trend Is Not a Natural Phenomenon

People are causing global warming by burning fossil fuels (like oil, coal and natural gas) and cutting down forests. Scientists have shown that these activities are pumping far more CO_2 into the atmosphere than was ever released in hundreds of thousands of years.

Environmental Defense Fund, "Global Warming Myths and Facts," 2008. www.edf.org/page. cfm?tagID = 1011.

Mixed Opinions on How to Proceed

But environmental advocates said the White House—which remains opposed to mandatory limits on U.S. carbon emissions—is making a mistake in assuming research and technological advances alone will address global warming.

"The administration's proposals are at least a decade away," said Angela Anderson, vice president for climate programs at the National Environmental Trust. "The promise of better technologies tomorrow shouldn't stop us from doing what we can today."

House and Senate Democratic leaders back a cap on greenhouse gases and hope to enact such legislation this year; [in February 2007], several of the report's authors are to testify in congressional hearings.

In an interview yesterday, House Science and Technology Committee Chairman Bart Gordon (D-Tenn.) called the report "a unanimous, definitive world statement"

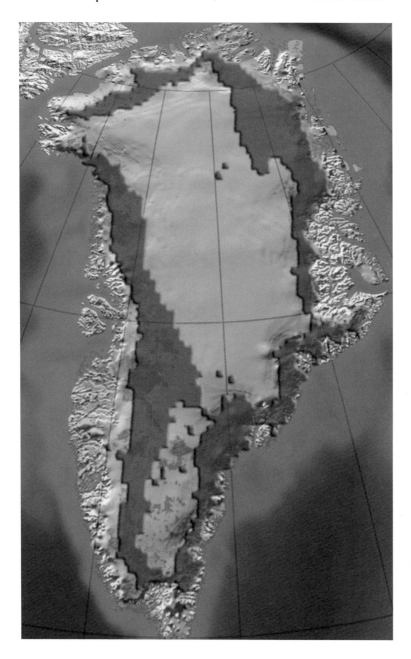

The orange areas on this map of Greenland show how much of Greenland's ice sheet has melted in recent years.

on climate change that, if anything, was too conservative. "It's time to end the debate and act," Gordon said. "All the naysayers should step aside."

Some critics, however, question the push for nation-wide limits on emissions from power plants, automobiles and other industrial sources. At the George C. Marshall Institute, a think tank that receives funding from Exxon Mobil, chief executive William O'Keefe and President Jeff Kueter issued a statement urging "great caution in reading too much" into the report until the panel releases its detailed scientific documentation a few months from now.

"Claims being made that a climate catastrophe later this century is more certain are unjustified," they said, adding that "the underlying state of knowledge does not justify scare tactics or provide sufficient support for proposals . . . to suppress energy use and impose large economic burdens on the U.S. economy."

Analyze the essay:

1. The subject of this essay is a report written by the IPCC on climate change. Explain in detail what the IPCC is, who wrote this report, and under what conditions. In your opinion, is the IPCC a credible source for people to use to make decisions about global warming? Why or why not?

2. The authors of the IPCC report said they are 90 percent certain that global warming is caused by humans. What do you think Dennis T. Avery, author of the following essay, would say in response to this claim? Quote from both texts in your answer.

Climate Change Is Not Necessarily Man-Made

Dennis T. Avery

In the following essay Dennis T. Avery argues that evidence is insufficient to conclude that global warming is a man-made phenomenon. He discusses scientific evidence that casts doubt on the theory that carbon dioxide (CO_2) from human industrial activity has caused the planet to warm. Avery points out that fossil records show the planet has endured warming periods centuries before human industrial activity began, and he argues that even contemporary warming periods might not be related to greenhouse gas emissions from human activity. Avery warns that claiming humans have caused climate change only encourages politicians to pass sweeping laws that would change American life as we know it. He argues that because we cannot be sure the planet is warming due to human activity, it is unethical to curb the economy and Americans' lifestyles because of it.

Avery is director of the Center for Global Food Issues at the Hudson Institute, a nonpartisan policy research organization dedicated to research and analysis in the areas of global security, prosperity, and freedom.

Consider the following questions:

1. According to Avery, what do long ice cores from Greenland and the Antarctic show about how often the planet undergoes warming periods?
2. What is the Pacific Decadal Oscillation, as reported by the author?
3. What do Antarctic ice cores show about the correlation between global warming and CO_2, according to Avery?

Dennis T. Avery, "Should We Believe the Latest UN Climate Report?" *Canada Free Press*, February 6, 2007. Reproduced by permission.

The UN Climate Change panel is asserting—again—that humans are overheating the planet. Again, they have no evidence to support their claim—but they want the U.S. to cut its energy use by perhaps 80 percent just in case. Stabilizing greenhouse gases means no personal cars, no air-conditioning, no vacation travel. [Democrat House majority leader] Nancy Pelosi says one-third of the Senate want this too.

It's a remarkably sweeping demand, given that the earth has warmed less than 1 degree C, over 150 years. This on a planet where the ice cores and seabed sediments tell us the climate has been either warming abruptly or cooling suddenly for the past million years.

A Long History of Warming and Cooling

The first long ice cores from Greenland and Antarctic were brought up in the 1980s. The ice layers showed the

Scientists harvest an ice core in the Arctic. The study of ice cores gives researchers data on the history of Earth's climate.

earth warming 1–2 degrees roughly every 1,500 years—usually suddenly. The natural warmings often gained half their total strength in a few decades, then waffled erratically for centuries—rather like our planet's temperature pattern since 1850.

History tells us the coolings, not the warmings, have been the bad part. After the Medieval Warming ended about 1300, Europe was hit by huge storms, gigantic sea floods, crop failures, and plagues of disease.

My big gripe with the IPCC [Intergovernmental Panel on Climate Change] is that they're still keeping this climate cycle a virtual secret from the public.

Preindustrial Warming

What does the IPCC say about hundreds of long-dead trees on California's Whitewing Mountain that tell us the earth was 3.2 degrees C warmer in the year 1350 than today? In that year, seven different tree species were killed—while growing above today's tree line—by a volcanic explosion. The trees' growth rings, species and location confirm that the climate was much warmer [than] that of today, says C. I. Millar of the U.S. Forest Service, reporting in *Quaternary Research*, Nov. 27, 2006.

The new IPCC report warns us it can't explain the recent surge of warming from 1976–1998. Therefore, it claims the surge must have been caused by human-emitted CO_2. But the IPCC also can't explain why more than half of the current warming occurred before 1940, before the Industrial Revolution improved global living standards and increased CO_2 emissions.

Look at this interesting coincidence: The "inexplicable" 1976–1998 surge in global temperature looks very much

> ## Hysteria over Climate Change
>
> When the new ice age predicted in the '70s failed to emerge, the eco-crowd moved on in the '80s to global warming, and then more recently to claiming as evidence of global warming every conceivable meteorological phenomenon: lack of global warmth is evidence of global warming; frost, ice, snow, glaciers, they're all signs of global warming, too. . . . Maybe we should all take a deep breath of CO_2 and calm down.
>
> Mark Steyn, "Climate Change Myth," *Australia*, September 11, 2006. www.freerepublic.com/focus/f-news/1555298/posts.

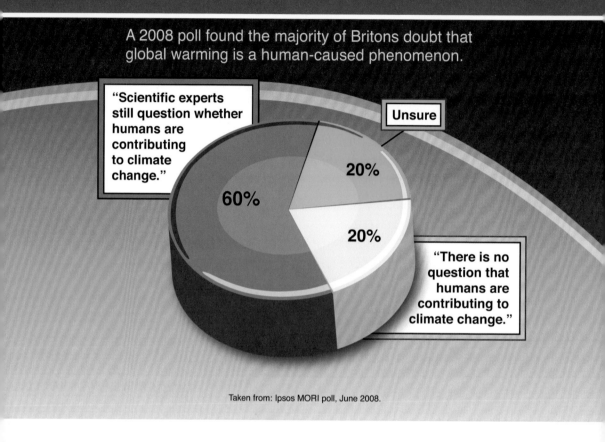

A 2008 poll found the majority of Britons doubt that global warming is a human-caused phenomenon.

"Scientific experts still question whether humans are contributing to climate change."

60%

Unsure

20%

20%

"There is no question that humans are contributing to climate change."

Taken from: Ipsos MORI poll, June 2008.

like the warming surge from 1916–1940. After 1940, we had a 35-year cooling—which the IPCC also can't explain. But in 1996, researchers discovered a 50–60 year Pacific-wide climate cycle they call the Pacific Decadal Oscillation [PDO]. This cycle caused the salmon decline in the Columbia River after 1977. It also causes shifts in sardine and anchovy catches all around the Pacific.

The PDO shifted into a cool phase in 1940, with lots of salmon in the Columbia, until 1977. That's almost exactly the period of the 1940–76 global cooling. Then the PDO turned warmer and the Columbia salmon declined—until about 1999. That closely matches the 1976–98 surge in global temperatures.

More Research Is Needed Before We Blame CO$_2$

Does the Pacific climate cycle explain the last two short-term blips on the world's temperature chart better than humanity's small contribution to the CO$_2$ that makes up only 0.03 percent of the atmosphere? It is certainly worth exploring more carefully before we make huge changes in our standards of living world-wide.

Past climate warmings haven't correlated with CO$_2$ changes. The Antarctic ice cores show that after the last four Ice Ages, the temperatures warmed 800 years before the CO$_2$ levels increased in the atmosphere. The warming produced more CO$_2$ in the atmosphere, not the other way around.

The author claims that the Pacific Decadal Oscillation effect, a 50–60-year Pacific climate cycle, shows that salmon populations declined in the Columbia River in periods of surges in global warming.

It's worth noting that the environmental movement and the politicians also blamed human activity for the salmon decline. Farming, fishing, and logging were reined in, sending the Pacific Northwest's rural economies into despair. Now we've found the PDO. Is a natural cycle also the answer for the UN climate change panel?

Analyze the essay:

1. Avery uses several pieces of evidence to support his claim that climate change is not necessarily man-made. List all the pieces of evidence Avery uses and then state how convincing you think each one is.

2. Avery states that it is unlikely humans have caused global warming because evidence shows that the earth has warmed prior to human industrial activity. After reading his essay and the one by Juliet Eilperin before it, what is your opinion on the role of humans in climate change?

Climate Change Will Fuel Regional Conflicts

German Advisory Council on Global Change

The following essay was written by the German Advisory Council on Global Change (WBGU), an independent, scientific advisory body established by the German government to analyze global environment and development problems. In the following report the WBGU warns that climate change will fuel regional conflicts and lead to global chaos. The authors say that climate change is expected to cause food and water shortages, damage regional economies, result in mass migrations that will overwhelm neighboring states, and have other devastating consequences that will wreak havoc on civilization. The more the climate changes, the more these problems will be exacerbated, they warn. They conclude that traditional security measures, such as government intervention and police and military squads, will not be enough to peacefully resolve the multitude of conflicts that are expected to result from climate change.

Consider the following questions:

1. What effect will water scarcity have on regional conflict, according to the authors?
2. What is the "equity gap," and what bearing does it have on the authors' argument?
3. In what way do the authors suspect climate change might affect human rights?

Climate Change as a Security Risk, London, UK: Earthscan with German Advisory Council on Global Change (WBGU), 2007. Copyright © 2008 German Advisory Council on Global Change. Reproduced by permission.

W ithout resolute counteraction, climate change will overstretch many societies' adaptive capacities within the coming decades. This could result in destabilization and violence, jeopardizing national and international security to a new degree. However, climate change could also unite the international community,

How Environmental Changes Impact Security and Conflict

It is believed that climate change will set off a series of events that will increase international violence and conflict.

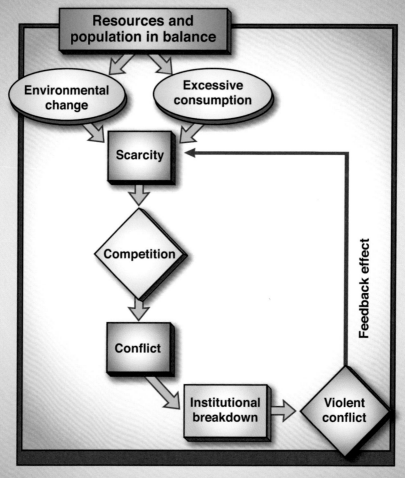

Taken from: B. Brown, A. Hammill, and R. McLeman, "Climate Change as the New Security Threat: Implications for Africa," *International Affairs*, vol. 83, no. 6, 2007.

provided that it recognizes climate change as a threat to humankind and soon sets the course for the avoidance of dangerous anthropogenic [caused by humans] climate change by adopting a dynamic and globally coordinated climate policy. If it fails to do so, climate change will draw ever-deeper lines of division and conflict in international relations, triggering numerous conflicts between and within countries over the distribution of resources, especially water and land, over the management of migration, or over compensation payments between the countries mainly responsible for climate change and those countries most affected by its destructive effects. . . .

Six Threats to International Stability and Security

In light of current knowledge about the social impacts of climate change, WBGU identifies the following six key threats to international security and stability which will arise if climate change mitigation fails:

1. *Possible increase in the number of weak and fragile states as a result of climate change:* Weak and fragile states have inadequate capacities to guarantee the core functions of the state, notably the state's monopoly on the use of force, and therefore already pose a major challenge for the international community. So far, however, the international community has failed to summon the political will or provide the necessary financial resources to support the long-term stabilization of these countries. Moreover, the impacts of unabated climate change would hit these countries especially hard, further limiting and eventually overstretching their problem-solving capacities. Conflict constellations may also be mutually reinforcing, e.g. if they extend beyond the directly affected region through environmental migration and thus destabilize other neighbouring states. This could ultimately lead to the emergence of "failing subregions" consisting of several simultaneously overstretched states, creating

"black holes" in world politics that are characterized by the collapse of law and public order, i.e. the pillars of security and stability. It is uncertain at present whether, against the backdrop of more intensive climate impacts, the international community would be able to curb this erosion process effectively.

A Crippled Global Economy

2. *Risks for global economic development:* Climate change will alter the conditions for regional production processes and supply infrastructures. Regional water scarcity will impede the development of irrigated agriculture and other water-intensive sectors. Drought and soil degradation will result in a drop in agricultural yields. More frequent extreme events such as storms and flooding put industrial sites and the transport, supply and production infrastructures in coastal regions at risk, forcing companies to relocate or close production sites. Depending on the type and intensity of the climate impacts, this could have a significant and adverse effect on the global economy. Unabated climate change is likely to result in substantially reduced rates of growth. This will increasingly limit the economic scope, at national and international level, to address the urgent challenges associated with the Millennium Development Goals [a set of eight goals developed by the United Nations that address the primary challenges faced by developing nations].

The Equity Gap

3. *Risks of growing international distributional conflicts between the main drivers of climate change and those most affected:* Climate change is mainly caused by the industrialized and newly industrializing countries. The major differences in the per capita emissions of industrialized and developing/newly industrializing countries are increasingly regarded as an "equity gap", especially as the rising costs of climate change are mainly being

borne by the developing countries. The greater the damage and the burden of adaptation in the South, the more intensive the distributional conflicts between the main drivers of climate change and those most affected will become. The worst affected countries are likely to invoke the "polluter pays" principle, so international controversy over a global compensation regime for climate change will probably intensify. Beside today's industrialized countries, the major ascendant economies whose emissions are increasing substantially, notably China but also India and Brazil, for example, will also be called to account by the developing countries in future. A key line of conflict in global politics in the 21st century would therefore divide not only the industrialized and the developing countries, but also the rapidly growing newly industrializing countries

Experts say that drought and resulting soil degradation due to climate change will result in lower crop yields, which will spark regional conflicts over food.

and the poorer developing countries. The international community is ill-prepared at present for this type of distributional conflict.

A Decline in Human Rights

4. *The risk to human rights and the industrialized countries' legitimacy as global governance actors:* Unabated climate change could threaten livelihoods, erode human security and thus contribute to the violation of human rights. Against the backdrop of rising temperatures, growing awareness of social climate impacts and inadequate climate change mitigation efforts, the CO_2-emitting industrialized countries and, in future, buoyant economies such as China could increasingly be accused of knowingly causing human rights violations, or at least doing so in *de facto* terms. The international human rights discourse in the United Nations is therefore also likely to focus in future on the threat that climate impacts pose to human rights. Unabated climate change could thus plunge the industrialized countries in particular into crises of legitimacy and limit their international scope for action.

The Challenges of Mass Migration

5. *Triggering and intensification of migration:* Migration is already a major and largely unresolved international policy challenge. Climate change and its social impacts will affect growing numbers of people, so the number of migration hotspots around the world will increase. The associated conflict potential is considerable, especially as "environmental migrants" are currently not provided for in international law. Disputes over compensation payments and the financing of systems to manage refugee crises will increase. In line with the "polluter pays" principle, the industrialized countries will have to face up to their responsibilities. If global temperatures continue to rise unabated, migration could become one of the major fields of conflict in international politics in future.

Old Security Methods Will Not Be Enough

6. *Overstretching of classic security policy:* The future social impacts of unabated climate change are unlikely to trigger "classic" interstate wars; instead, they will probably lead to an increase in destabilization processes and state failure with diffuse conflict structures and security threats in politically and economically overstretched states and societies. The specific conflict constellations, the failure of disaster management systems after extreme weather events and increasing environmental migration will be almost impossible

Climate Change Will Cause Conflict

Researchers predict that climate change will cause food and water shortages, mass migration, and disastrous weather conditions, which will result in increased regional conflict.

 Climate-induced degradation of freshwater resources

 Climate-induced decline in food production

 Conflict hotspot

 Climate-induced increase in storm and flood disasters

 Environmentally-induced migration

Taken from: German Advisory Council on Global Change, 2007.

to manage without support from police and military capacities, and therefore pose a challenge to classic security policy. In this context, a well-functioning cooperation between development and security policy will be crucial, as civilian conflict management and reconstruction assistance are reliant on a minimum level of security. At the same time, the largely unsuccessful operations by highly equipped military contingents which have aimed to stabilize and bring peace to weak and fragile states since the 1990s show that "classic" security policy's capacities to act are limited. A climate-induced increase in the number of weak and fragile states or even the destabilization of entire subregions would therefore overstretch conventional security policy.

How Climate Change Causes Regional Instability

Floods in the Ganges caused by melting glaciers in the Himalayas are wreaking havoc in Bangladesh leading to a rise in illegal migration to India. This has prompted India to build an immense border fence in attempt to block newcomers. Some 6,000 people illegally cross the border to India every day.

John Reid, "Water Wars: Climate Change May Spark Conflict," *Independent* (UK), February 28, 2006. www.independent.co.uk/environment/water-wars-climate-change-may-spark-conflict-467957.html.

More Insecurity with Each Degree

The greater the scale of climate change, the greater the probability that in the coming decades, climate-induced conflict constellations will impact not only on individual countries or subregions but also on the global governance system as a whole. These new global risk potentials can only be countered by policies that aim to manage global change. Every one of the six threats to international stability and security, outlined above, is itself hard to manage. The interaction between these threats intensifies the challenges for international politics. It is almost inconceivable that in the coming years, a global governance system could emerge with the capacity to respond effectively to the conflict constellation identified by WBGU. Against the backdrop of globalization, unabated climate change

is likely to overstretch the capacities of a still insufficient global governance system.

As the climate-induced security risks of the 21st century have their own specific characteristics, they will be difficult to mitigate through classic military interventions. Instead, an intelligent and well-crafted global governance strategy to mitigate these new security risks would initially consist of an effective climate policy, which would then evolve into a core element of preventive security policy in the coming decades. The more climate change advances, the more important adaptation strategies in the affected countries will become, and these must be supported by international development policy. At international level, the focus will be on global diplomacy to contain climate-induced conflicts, as well as on the development of compensation mechanisms for those affected by climate change, global migration

Afghan refugees flee their villages due to drought. As more droughts occur because of global warming, more regional conflicts will be ignited.

policy, and measures to stabilize the world economy. The opportunities to establish a well-functioning global governance architecture will narrow as global temperatures rise, revealing a vicious circle: climate change can only be combated effectively through international cooperation, but with advancing climate change, the basis for constructive multilateralism will diminish. Climate change thus poses a challenge to international security, but classic, military-based security policy will be unable to make any major contributions to resolving the impending climate crises.

Analyze the essay:

1. In this essay WBGU authors use logical reasoning, examples, and history to make their argument that climate change will contribute to global insecurity. They do not, however, use any quotations to support their points. If you were to rewrite this essay and insert quotations, what authorities might you quote from? Where would you place these quotations to bolster the points the authors make?

2. The authors of this essay argue that climate change will cause global conflict. Idean Salehyan, author of the following essay, disagrees. After reading both essays, what is your opinion on the effect of climate change on global security? Use evidence from the texts in your answer.

Climate Change Will Not Fuel Regional Conflicts

Idean Salehyan

In the following essay Idean Salehyan challenges an emerging theory that global warming will create regional instability and lead to global conflict. Salehyan explains that historically, an abundance of, rather than a lack of, resources has sparked regional conflict. Furthermore, as the climate has continued to warm, regional conflicts and civil wars have been on the decline. The author points out that environmental crises such as the 2004 Asian tsunami did not lead to increased regional conflict and argues there is no reason to think that other climate change–related disasters would be any different. In Salehyan's opinion, blaming climate change for conflict lets corrupt, violent governments off the hook for stirring up global instability and invites the military to control problems that would be better solved by technological and agricultural innovations. For all of these reasons Salehyan concludes that spreading the idea that climate change causes global instability is not only wrong, but also irresponsible.

Salehyan is assistant professor of political science at the University of North Texas and coauthor of *Climate Change and Conflict: The Migration Link.*

Consider the following questions:

1. What did data collected by Uppsala University and the International Peace Research Institute show about the state of armed conflicts and global warming, as reported by the author?
2. How does the author use the nation of Malawi to support his argument?
3. In the author's opinion, what is a likely cause of violence in Sudan?

Idean Salehyan, "The New Myth About Climate Change," *Foreign Policy*, August 2007. Reproduced by permission.

Few serious individuals still contest that global climate change is among the most important challenges of our time. The overwhelming scientific consensus is that global warming is a very real phenomenon, that human activity has contributed to it, and that some degree of climate change is inevitable.

We are no longer arguing over the reality of climate change, but rather, its potential consequences. According to one emerging "conventional wisdom," climate change will lead to international and civil wars, a rise in the number of failed states, terrorism, crime, and a stampede of migration toward developed countries.

The New Myth About Climate Change

It sounds apocalyptic, but the people pushing this case are hardly a lunatic fringe. United Nations Secretary-General Ban Ki-moon, for instance, has pointed to climate change as the root cause of the conflict in Darfur. A group of high-ranking retired U.S. military officers recently published a report that calls climate change "a threat multiplier for instability." An earlier report commissioned by the Pentagon argues that conflicts over scarce resources will quickly become the dominant form of political violence. Even the Central Intelligence Agency is reportedly working on a National Intelligence Estimate that will focus on the link between climate change and U.S. national security.

These claims generally boil down to an argument about resource scarcity. Desertification, sea-level rise, more-frequent severe weather events, an increased geographical range of tropical disease, and shortages of freshwater will lead to violence over scarce necessities. Friction between haves and have-nots will increase, and gov-

---

Scarce Resources Lead to Cooperation, Not Conflict

Historically, water scarcity has often—though certainly not always—worked to favour cooperation between states. Interstate dialogue prompted by diminished water supplies, particularly, can build trust, institutionalise cooperation on a broader range of issues and create common regional identities.

International Crisis Group, "Climate Change and Conflict," August 2008. www.crisisgroup.org/home/index.cfm?id = 4932.

ernments will be hard-pressed to provide even the most basic services. In some scenarios, mass migration will ensue, whether due to desertification, natural disasters, and rising sea levels, or as a consequence of resource wars. Environmental refugees will in turn spark political violence in receiving areas, and countries in the "global North" will erect ever higher barriers to keep culturally unwelcome—and hungry—foreigners out.

Extreme drought conditions in the African nation of Malawi have not brought on a civil war despite acute food shortages.

The number of failed states, meanwhile, will increase as governments collapse in the face of resource wars and weakened state capabilities, and transnational terrorists and criminal networks will move in. International wars over depleted water and energy supplies will also intensify. The basic need for survival will supplant nationalism, religion, or ideology as the fundamental root of conflict.

These Are Misleading Scenarios

Dire scenarios like these may sound convincing, but they are misleading. Even worse, they are irresponsible, for they shift liability for wars and human rights abuses away from oppressive, corrupt governments. Additionally, focusing on climate change as a security threat that requires a military response diverts attention away from prudent adaptation mechanisms and new technologies that can prevent the worst catastrophes.

First, aside from a few anecdotes, there is little systematic empirical evidence that resource scarcity and changing environmental conditions lead to conflict. In fact, several studies have shown that an *abundance* of natural resources is more likely to contribute to conflict. Moreover, even as the planet has warmed, the number of civil wars and insurgencies has decreased dramatically. Data collected by researchers at Uppsala University and the International Peace Research Institute, Oslo shows a steep decline in the number of armed conflicts around the world. Between 1989 and 2002, some 100 armed conflicts came to an end, including the wars in Mozambique, Nicaragua, and Cambodia. If global warming causes conflict, we should not be witnessing this downward trend.

Environmental Catastrophes Do Not Lead to Conflict

Furthermore, if famine and drought led to the crisis in Darfur, why have scores of environmental catastrophes failed to set off armed conflict elsewhere? For instance, the U.N. World Food Programme warns that 5 million people in Malawi have been experiencing chronic food shortages for several years. But famine-wracked Malawi has yet to experience a major civil war. Similarly, the Asian tsunami in 2004 killed hundreds of thousands of people, generated millions of environmental refugees, and led to severe shortages of shelter, food, clean water, and electricity. Yet the tsunami, one of the most extreme catastrophes in recent history, did not lead to an outbreak of resource wars. Clearly then, there is much more to armed conflict than resource scarcity and natural disasters.

Second, arguing that climate change is a root cause of conflict lets tyrannical governments off the hook. If the environment drives conflict, then governments bear little responsibility for bad outcomes. That's why Ban Ki-moon's case about Darfur was music to Khartoum's ears. The Sudanese government would love to blame the

Climate Change Can Spark Cooperation Rather than Conflict

Historically, resource crises such as water scarcity have encouraged nation-states to work together for their common benefit. Africa is one place where governments have worked together to protect natural areas and conserve resources by building Peace Parks.

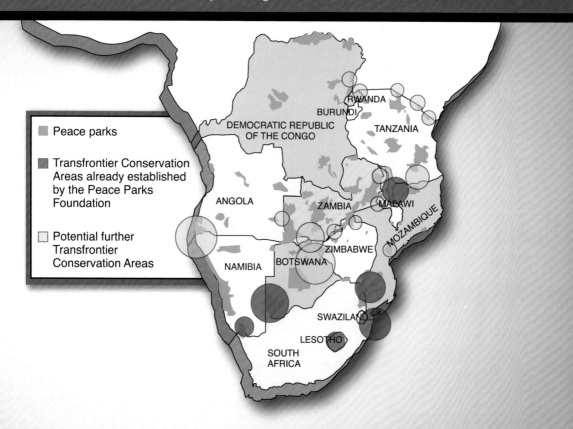

Taken from: Alexander Carius, "Environmental Peacebuilding: Conditions for Success," Woodrow Wilson International Center for Scholars, no.12, 2006–2007, p. 69.

Sudanese drought refugees wait outside an aid station for relief. The author claims that Sudan's problems are more the fault of the Sudanese government than climate change.

West for creating the climate change problem in the first place. True, desertification is a serious concern, but it's preposterous to suggest that poor rainfall—rather than deliberate actions taken by the Sudanese government and the various combatant factions—ultimately caused the genocidal violence in Sudan. Yet by Moon's perverse logic, consumers in Chicago and Paris are at least as culpable for Darfur as the regime in Khartoum.

Corrupt Governments Lead to Conflict

To be sure, resource scarcity and environmental degradation can lead to social frictions. Responsible, accountable governments, however, can prevent local squabbles from spiraling into broader violence, while mitigating the risk of some severe environmental calamities. As Nobel laureate Amartya Sen has observed, no democracy has ever experienced a famine. Politicians who fear the wrath of voters usually do their utmost to prevent foreseeable disasters and food shortages. Accountable leaders are also better at providing public goods such as clean air and water to their citizens.

Third, dire predictions about the coming environmental wars imply that climate change requires military solutions—a readiness to forcibly secure one's own resources, prevent conflict spillovers, and perhaps gain control of additional resources. But focusing on a military response diverts attention from simpler, and far cheaper, adaptation mechanisms. Technological improvements in agriculture, which have yet to make their way to many poor farmers, have dramatically increased food output in the United States without significantly raising the amount of land under cultivation. Sharing simple technologies with developing countries, such as improved irrigation techniques and better seeds and fertilizers, along with finding alternative energy supplies and new freshwater sources, is likely to be far more effective and cost saving in the long run than arms and fortifications. States affected by climate change can move people out of flood plains and desert areas, promote better urban planning, and adopt more efficient resource-management systems.

Conflict Is Not a Consequence of Climate Change

Yes, climate change is a serious problem that must be addressed, and unchecked environmental degradation may lead to intensified competition over scarce resources in certain regions. The good news is that the future is not written in stone. How governments respond to the challenge is at least as important as climate change itself, if not more so. Well-managed, transparent political systems that are accountable to their publics can take appropriate measures to prevent armed conflict. If the grimmest scenarios come to pass and environmental change contributes to war, human rights abuse, and even genocide, it will be reckless political leaders who deserve much of the blame.

Analyze the essay:

1. In this essay Salehyan argues that historically, regional conflict and wars have been started over an abundance of resources, rather than a lack of them. How do you think the author of the preceding viewpoint, the German Advisory Council on Global Change, might respond to this argument? Explain your answer using evidence from the texts.

2. Salehyan suggests that as the climate has warmed, global conflict has declined. What pieces of evidence did he provide to support this claim? Did he convince you of his argument? Explain why or why not.

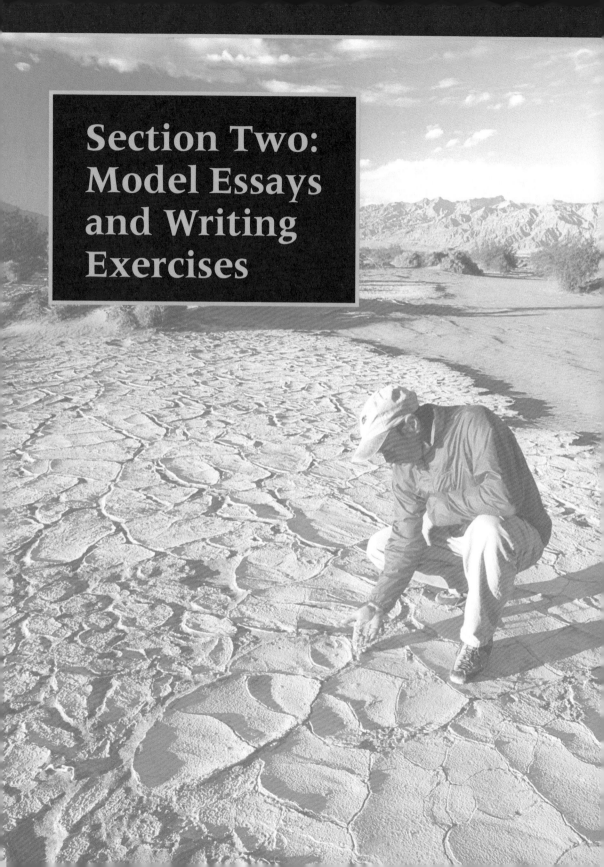

**Section Two:
Model Essays
and Writing
Exercises**

The Five-Paragraph Essay

An *essay* is a short piece of writing that discusses or analyzes one topic. The five-paragraph essay is a form commonly used in school assignments and tests. Every five-paragraph essay begins with an *introduction*, ends with a *conclusion*, and features three *supporting paragraphs* in the middle.

The Thesis Statement. The introduction includes the essay's thesis statement. The thesis statement presents the argument or point the author is trying to make about the topic. The essays in this book all have different thesis statements because they are making different arguments about climate change.

The thesis statement should clearly tell the reader what the essay will be about. A focused thesis statement helps determine what will be in the essay; the subsequent paragraphs are spent developing and supporting its argument.

The Introduction. In addition to presenting the thesis statement, a well-written introductory paragraph captures the attention of the reader and explains why the topic being explored is important. It may provide the reader with background information on the subject matter or feature an anecdote that illustrates a point relevant to the topic. It could also present startling information that clarifies the point of the essay or put forth a contradictory position that the essay will refute. Further techniques for writing an introduction are found later in this section.

The Supporting Paragraphs. The introduction is then followed by three (or more) supporting paragraphs. These are the main body of the essay. Each paragraph presents and develops a *subtopic* that supports the essay's thesis statement. Each subtopic is spearheaded by a *topic sentence* and supported by its own facts, details, and examples. The writer can use various kinds of supporting material and details to back up the topic of each supporting paragraph. These may include statistics, quotations from people with special knowledge or expertise, historic facts, and anecdotes. A rule of writing is that specific and concrete examples are more convincing than vague, general, or unsupported assertions.

The Conclusion. The conclusion is the paragraph that closes the essay. Its function is to summarize or reiterate the main idea of the essay. It may recall an idea from the introduction or briefly examine the larger implications of the thesis. Because the conclusion is also the last chance a writer has to make an impression on the reader, it is important that it not simply repeat what has been presented elsewhere in the essay but close it in a clear, final, and memorable way.

Although the order of the essay's component paragraphs is important, they do not have to be written in the order presented here. Some writers like to decide on a thesis and write the introduction paragraph first. Other writers like to focus first on the body of the essay and write the introduction and conclusion later.

Pitfalls to Avoid

When writing essays about controversial issues such as climate change, it is important to remember that disputes over the material are common precisely because there are many different perspectives. Remember to state your arguments in careful and measured terms. Evaluate your topic fairly—avoid overstating negative qualities of one perspective or understating positive qualities of another. Use examples, facts, and details to support any assertions you make.

The Compare-and-Contrast Essay

The last section of this book provided you with samples of previously published writing on climate change. All are persuasive, or opinion, essays that make certain arguments about various topics relating to climate change. Most of them also compare and contrast different facts and figures to make their arguments. This section will focus on writing your own compare-and-contrast essay.

In terms of presenting information and making an argument, the compare-and-contrast method is a very effective way to organize an essay. At the heart of the compare-and-contrast essay is the act of evaluating two or more issues, things, or ideas next to each other. Such side-by-side glances can often reveal aspects of one subject that might have gone unnoticed had it been evaluated by itself.

Ways to Structure Compare-and-Contrast Essays

There are two basic ways to structure your compare-and-contrast essay. You can either evaluate your subjects point by point, analyzing them together throughout the essay. Or, you can evaluate your subjects separately, reserving the first half of the essay for one subject and the second half of the essay for the other. Sometimes you may find it is effective to mix these two approaches, but in general the form you choose will determine the overall structure, pacing, and flow of the essay.

Sometimes, the compare-and-contrast essay can focus on either the similarities *or* the differences between your subjects. A comparison essay usually explores the similarities between two subjects, while a contrast essay focuses on their differences. You can solely compare your subjects in order to present their commonality. Or, you can solely contrast your subjects in order to expose their fundamental differences. Of course, you can also compare and contrast in the same essay. This approach is useful when your subjects are not entirely similar or different, which is often the case. This method can be effective when attempting to define, analyze, or arrive at a more in-depth understanding of your subjects.

Compare-and-contrast essays can also have one of two purposes. They can be used to conduct objective, unbiased discussions of two or more subjects. Or, they can be useful for making persuasive arguments in which you attempt to convince the reader of something. The effect can be achieved by evaluating two subjects' advantages and disadvantages or showing in what ways one is superior to another. Then you would advocate a course of action or express a preference for one over the other.

Tips to Remember

Regardless of what style of compare-and-contrast essay you choose, certain features are common to all of them. For example, all compare-and-contrast essays focus on at least two subjects. Furthermore, all compare-and-contrast essays feature certain transitional words that signal a similarity or difference is being pointed out.

When writing compare-and-contrast essays it is important to choose two subjects that are comparable or contrastable. For example, you could write a simple compare-and-contrast essay that focuses on the similarities and differences of oranges and apples or T-shirts and sweaters. You would not, however, want to set out to compare oranges and tables or T-shirts and rice. The subjects you choose to compare or contrast must be linked in a basic way so that they warrant an examination. In terms of climate change, it makes sense to compare and contrast different scientific methods for determining whether climate change is occurring. It also makes sense to compare and contrast different suggestions for curbing climate change. It would not, however, make sense to compare solutions for curbing climate change with solutions for preserving endangered species, because they are inherently different subjects.

In the following section you will read some model essays on climate change that use compare-and-contrast arguments. You will also complete exercises that will help you write your own.

Words and Phrases Common in Compare-and-Contrast Essays

additionally	in comparison
also; too	in contrast
alternatively	in the same way
as well as	likewise
conversely	moreover
equally	on the contrary
from this perspective	on the other hand
furthermore	similarly
however	then again

Is the Earth Warming or Cooling?

| Editor's Notes | A compare-and-contrast essay can be written in several different ways. One way is to analyze two subjects' similarities or differences. Another is to describe thing or argument A, describe thing or argument B, and then compare them. The following five-paragraph essay does just that: It compares theories on whether the planet is warming or cooling. The essay is structured as a five-paragraph essay in which each paragraph contributes a supporting piece of evidence to develop the argument.

The notes in the margin point out key features of the essay and will help you understand how the essay is organized. Also note that all sources are cited using Modern Language Association (MLA) style.* For more information on how to cite your sources, see Appendix C. In addition, consider the following:

1. How does the introduction engage the reader's attention?
2. What pieces of supporting evidence are used to back up the essay's points and arguments?
3. What purpose do the essay's quotations serve?
4. How does the author transition from one idea to another?
5. What compare-contrast techniques are used?

Paragraph 1

Though climate change is an increasingly hot topic, it remains poorly understood by politicians, journalists, everyday citizens, and even scientists. In fact, so much confusion exists about climate change that people cannot

---

* Editor's Note: In applying MLA style guidelines in this book, the following simplifications have been made: Parenthetical text citations are confined to direct quotations only; electronic source documentation in the Works Cited list omits date of access, page ranges, and some detailed facts of publication.

reach agreement on whether climate change is occurring, what effect it might have on crops or sea levels, and whether it is man-made. Even when people agree that climate change is occurring, they disagree on whether that change is heralding a warmer planet or a cooler one. Heated debates over whether the planet is warming or cooling serve to illustrate the extent to which climate change is a complex topic.

This is the essay's thesis statement. It tells the reader what will be discussed in the following paragraph.

Paragraph 2

This is the topic sentence of paragraph 2. It is a subset of the essay's thesis. It tells what specific point this paragraph will make.

The better documented, and certainly oft-repeated, theory is that the planet is warming, and at an alarming rate. According to scientists Peter Backlund, Anthony Janetos, and David Schimel, authors of a 2008 paper on climate change published by the U.S. Department of Agriculture, the global average surface temperature during the twentieth century increased by about 1.08°F (0.6°C), an increase that, if continued, could spell disaster for the planet's ecosystems. Furthermore, warmer temperatures caused glaciers to melt, releasing water into the seas. This is how scientists explain a global sea level increase of about 6 to 8 inches (15 to 20 cm). Lee Dye, a science writer who lives in Alaska, reports that the glaciers around his home are visibly melting from a warmer climate: "The Mendenhall [glacier] is rapidly becoming a shadow of its former self. It is melting and receding at a rate of several hundred feet a year," he says. "Just 200 years ago, the toe of the glacier was where the Juneau airport is today. Now it's several miles—that's miles—back into the spruce-covered hills." This is just one way in which the planet's decades-long warming trend has become apparent.

The author lists the credentials of the people she cites to prove that they are knowledgeable. Always cite sources that can speak credibly on your topic.

Note how the author uses this account to support the essay's thesis and to personalize the problem of climate change.

Paragraph 3

Yet on the other hand, evidence exists that the planet is cooling. For example, the four agencies that track the planet's temperature—the NASA Goddard Institute for Space Studies, the Hadley Climate Research Unit, the Christy group at the University of Alabama, and Remote Sensing Systems, Inc.—report that in 2007 Earth's tem-

"On the other hand" is a transitional phrase common to compare-contrast essays. It keeps the sentences linked together and keeps ideas moving. For more phrases, see Preface B.

perature actually cooled by about 1.3°F (0.7°C). Such a cooling had not been recorded since 1930. "Disconcerting as it may be to true believers in global warming," says geophysicist Phil Chapman, "the average temperature on Earth has remained steady or slowly declined during the past decade, despite the continued increase in the atmospheric concentration of carbon dioxide, and now the global temperature is falling precipitously." Chapman further proves that 2007 was a surprisingly cold year by including the fact that winter brought snow to Baghdad and that the Antarctic accumulated more ice than it had had since 1770.

> This is a *supporting detail*. This information directly supports the paragraph's topic sentence, helping to prove it true.

Paragraph 4

So which is it—is the planet warming or cooling? One way to answer that question is to compare the methods used for evaluating the planet's average temperature. For example, the above evidence presented by Chapman is largely based on the year 2007—yet it is difficult to make projections about climate change, a process which occurs slowly over long periods of time, based on data from just one year. Conversely, that same logic can be applied to theories that argue the planet is warming— for example, a common criticism of warming theories is that they discount the fact that the planet has endured several periods of warming in its history, many of which were more extreme than that which is occurring today. In other words, both theories can be called into question by claiming that researchers are not looking broadly enough at their data.

> This is another transitional phrase common to compare-contrast essays.

> This is the sentence that best expresses the main idea of paragraph 4. Topic sentences don't always appear first in the paragraph.

Paragraph 5

But either way, it appears that both a cooler or warmer planet could spell trouble for our environment. Both extreme cold and extreme heat could be disastrous for crops—a warmer one helping pests and disease to thrive, a colder one limiting the growing season and available land for agriculture. A cooler or warmer planet would also threaten swaths of the Earth's population, a warmer

> The author compares the disastrous similarities of a warmer and cooler climate.

one threatening people in island nations and a cooler one threatening those who have settled in ice-age-prone North America and northern Europe. Indeed, perhaps the most critical aspect to remember about debates over climate change is that both extreme warming and extreme cooling could have severe consequences for the planet and all of its inhabitants. Scientists must continue to research whether the planet is undergoing a warming or a cooling trend so governments of all nations are as prepared as possible to respond.

Works Cited

Chapman, Phil. "Sorry to Ruin the Fun, but an Ice Age Cometh." *The Australian* 23 Apr. 2008 < http://www. theaustralian.news.com.au/story/0,25197,235833 76-7583,00.html >.

Dye, Lee. "Global Climate Change Is Happening Now: Scientists Fear Global Warming Is Irreversible and Its Effects Possibly Disastrous." *ABC News* 12 July 2006 < http://abcnews.go.com/Technology/Story?id = 2182824&page = 1 >.

Exercise 1A: Create an Outline from an Existing Essay

It often helps to create an outline of the five-paragraph essay before you write it. The outline can help you organize the information, arguments, and evidence you have gathered during your research.

For this exercise, create an outline that could have been used to write *Is the Earth Warming or Cooling?* This "reverse engineering" exercise is meant to help familiarize you with how outlines can help classify and arrange information.

To do this you will need to

1. articulate the essay's thesis,
2. pinpoint important pieces of evidence,
3. flag quotations that support the essay's ideas, and
4. identify key points that support the argument.

Part of the outline has already been started to give you an idea of the assignment.

Outline

I. Paragraph 1
Write the essay's thesis or main objective: to compare theories on whether the planet is warming or cooling.

II. Paragraph 2
Topic:

   Supporting Detail i: The global average surface temperature during the twentieth century increased by about 0.6°C.

   Supporting Detail ii:

III. Paragraph 3
Topic: Evidence exists that the planet is cooling.

    i.

    ii.

IV. Paragraph 4
Topic: Both theories can be called into question by claiming that researchers are not looking broadly enough at their data.

    i.

    ii.

V. Paragraph 5
Write the essay's conclusion:

Climate Change Spells Disaster for Crops

Editor's Notes The second essay, also written in five paragraphs, is a slightly different type of compare-and-contrast essay from the first model essay. In the first essay, the author contrasted the differences between two subjects without expressing a perspective or opinion. In the following essay the author compares two subjects and expresses a preference for one side over another. This is called a persuasive or opinionated essay and is meant to convince the reader to agree with the author's point of view.

The notes in the sidebars provide questions that will help you analyze how this essay is organized and how it is written.

Paragraph 1

Global warming promises to bring many significant changes for the environment, and thus for civilization. Yet it remains to be seen whether these changes would help or hurt humanity. How climate change might affect agriculture is a chief concern in this area—severe changes to crops could trigger a global famine and mass food shortages. Yet it has also been suggested that climate change might benefit the world's hungry populations by increasing crop yields and lengthening growing seasons. While this optimistic view of climate change would be nice, a comparison of each side's claims reveals that a warmer climate is more likely to diminish global food supplies than increase them.

What is the essay's thesis statement? How did you recognize it? What indicates this will be a compare-and-contrast essay?

Paragraph 2

Those who believe climate change would benefit crops often claim that a warmer climate is likely to result in longer growing seasons. While this development might seem to offer more time for crops to yield their fruit, it is more

What is the topic sentence of paragraph 2? Look for a sentence that tells generally what the paragraph's main point is.

likely that these long warm periods would be accompanied by severe weather, such as hurricanes, droughts, and floods that would mitigate the benefit of an extended season. Furthermore, while some crops would flourish in temperatures brought on by a warmer spring and fall, it is likely that the peak temperatures reached in the summer would be too hot. For this reason, researchers at the German Advisory Council for Global Change, a climate change research institute, predict that "a temperature rise of just 2°C [could cause] a drop in agricultural productivity . . . worldwide." In addition, warmer temperatures reduce some plants' ability to fertilize their seeds, which would further diminish crop yields.

> "For this reason" is just one transitional phrase that appears in this essay. Can you find all of them? Make a list.

Paragraph 3

Another claim is that many plants thrive when they are exposed to increased levels of carbon dioxide (CO_2), a key feature of global warming. Indeed, high exposure to CO_2 is believed to actually help plants grow faster and bigger. According to Dennis T. Avery, director of global food issues at the Hudson Institute, and H. Sterling Burnett, a senior fellow at the National Center for Policy Analysis, "A doubling of CO_2 from present levels would improve plant productivity on average by 32 percent across species." But it is important to note exactly what kinds of plants thrive well under these conditions. While hardy grains such as rice and wheat can thrive in warmer temperatures and higher levels of CO_2, scientists Peter Backlund, Anthony Janetos, and David Schimel report that more sensitive crops, such as tomatoes, onions, and fruits, do not do well. Furthermore, one plant species that reportedly responds fantastically to increased levels of CO_2 is weeds. If weeds flourish, they threaten to take over more valuable, edible crops. In addition, Backlund, Janetos, and Schimel report that herbicides—chemicals used in farming to kill weeds so crops can thrive—are less effective against CO_2-rich weeds. Clearly, increased CO_2 levels would not benefit the kinds of plants that communities depend on for food.

> Make a list of everyone quoted in this essay. What types of people have been quoted? What makes them qualified to speak on this topic?

> Identify a piece of evidence used to support paragraph 3's main idea.

Paragraph 4

Finally, a warmer climate is often touted as being able to reduce the severity of crop-killing frosts that come with colder weather—unfortunately, it is these frosts that often help keep pests and disease in check. "Disease pressure on crops and domestic animals will likely increase with earlier springs and warmer winters, which will allow proliferation and higher survival rates of pathogens and parasites." (Backlund, Janetos, and Schimel 2005, 6) This is yet another way in which climate change is expected to severely reduce harvests and food supplies and eventually herald a worldwide hunger crisis.

How is the topic of paragraph 4 different from, but related to, the other topics discussed thus far?

What point in paragraph 4 does this quotation support?

Paragraph 5

While no one knows for sure what effect climate change would have on agriculture, it seems likely that even a small temperature change of a degree or two would have disastrous consequences for the crops we rely on for food. Scientists, politicians, and researchers must continue to look into how climate change would affect crops and come up with contingency plans for averting a food crisis before it is too late.

Note how the author returns to ideas introduced in paragraph 1. See Exercise 3A for more on introductions and conclusions.

Works Cited

Avery, Dennis T., and H. Sterling Burnett. "Global Warming: Famine—or Feast?" Brief Analysis no. 517. National Center for Policy Analysis. 19 May 2005 < http://www.ncpa.org/pub/ba/ba517/ > .

Backlund, Peter, Anthony Janetos, and David Schimel. "The Effects of Climate Change on Agriculture, Land Resources, Water Resources, and Biodiversity: Executive Summary." U.S. Department of Agriculture. May 2008 < http://www.usda.gov/oce/global_change/files/SAP 4_3/ExecSummary.pdf > .

German Advisory Council on Global Change, 2007, "Climate Change as a Security Risk." < http://www.wbgu.de /wbgu_jg2007_engl.pdf > .

Exercise 2A: Create an Outline from an Existing Essay

As you did for the first model essay in this section, create an outline that could have been used to write *Climate Change Spells Disaster for Crops*. Be sure to identify the essay's thesis statement, its supporting ideas, its descriptive passages, and key pieces of evidence that were used.

Exercise 2B: Create an Outline for Your Own Essay

The second model essay expresses a particular point of view about climate change. For this exercise, your assignment is to find supporting ideas, choose specific and concrete details, create an outline, and ultimately write a five-paragraph essay making a different, or even opposing, point about climate change. Your goal is to use compare-and-contrast techniques to convince your reader.

Part l: Write a thesis statement.

The following thesis statement would be appropriate for an opposing essay on why climate change is likely to improve crop yields:

> *Because climate change is likely to bring warmer weather, abundant rainfall, and a CO_2-rich environment, crop yields and food production are likely to soar.*

Or see the sample paper topics suggested in Appendix D for more ideas.

Part II: Brainstorm pieces of supporting evidence

Using information from some of the viewpoints in the previous section and from the information found in Section Three of this book, write down three arguments or pieces of evidence that support the thesis statement you selected. Then, for each of these three arguments, write down supportive facts, examples, and details that support it. These could be:

- statistical information;
- the findings of reputable research papers;
- opinionated quotations from experts, peers, or family members;
- observations of people's actions and behaviors;
- specific and concrete details;
- critical analysis of the way information was gathered or presented.

Supporting pieces of evidence for the above sample topic sentence are found in this book and include:

- The quotation box accompanying Viewpoint Two from Dennis T. Avery and H. Sterling Burnett claiming that increased CO_2 levels have increased food production worldwide.
- Data presented in Viewpoint Two that challenge the assumption that climate change will occur or that it will be severe.
- Claims made in Viewpoint Four that show that historically, cooling periods have brought more disaster than warming periods.
- Point made by Idean Salehyan in Viewpoint Six that historically, abundance, rather than scarcity, has fueled global conflict.

Part III: Place the information from Parts I and II in outline form.

Part IV: Write the arguments or supporting statements in paragraph form.

By now you have three arguments that support the paragraph's thesis statement, as well as supporting material. Use the outline to write out your three supporting arguments in paragraph form. Make sure each paragraph has a topic sentence that states the paragraph's thesis clearly and broadly. Then, add supporting sentences that express the facts, quotations, details, and examples that support the paragraph's argument. The paragraph may also have a concluding or summary sentence.

Can Renewable Energy Help Avert Climate Change?

Editor's Notes The final model essay explores the pros and cons of using renewable sources of energy to avoid climate change. It lays out the benefits and drawbacks of these sources and concludes that America's energy policy should incorporate, but not entirely depend on, renewable resources.

This essay differs from the previous model essays in that it is longer than five paragraphs. Sometimes five paragraphs are simply not enough to adequately develop an idea. Extending the length of an essay can allow the reader to explore a topic in more depth or present multiple pieces of evidence that together provide a complete picture of a topic. Longer essays can also help readers discover the complexity of a subject by examining a topic beyond its superficial exterior. Moreover, the ability to write a sustained research or position paper is a valuable skill you will need as you advance academically.

As you read, consider the questions posed in the margins. Continue to identify thesis statements, supporting details, transitions, and quotations. Examine the introductory and concluding paragraphs to understand how they give shape to the essay. Finally, evaluate the essay's general structure and assess its overall effectiveness.

Paragraph 1

It is often said that when it comes to energy, there is no "free lunch"—that is, it is impossible to power modern society without affecting the environment in some way. If so, our current reliance on fossil fuels is a very expensive meal indeed. When oil is consumed it produces various pollutants, which include smog, tailings, and other muck that pollutes the air, water, and land. But of course, oil's most dangerous by-product is carbon dioxide, which is

82

believed to be responsible for climate change. When fossil fuels are burned, they produce gases such as carbon dioxide that collect in Earth's atmosphere. When too much of these gases are present, they trap heat in the atmosphere, causing the planet to warm. Over time, these slight increases in temperature can have disastrous environmental consequences. Delicate ice caps can melt and cause sea levels to rise, which in turn threaten to cover low-lying islands, erode shores, and destroy marine life. Furthermore, climate change can cause an increase in disease; trigger severe weather such as tornados, earthquakes, and hurricanes; and threaten the normal growth of food crops. These are just some of the ways in which the burning of fossil fuels heralds climate change.

How does the introduction set the stage for the topic that will be discussed?

Paragraph 2

Given this reality, it is imperative that humanity begin seeking power sources that have the least environmental impact possible. It is often suggested that renewable resources—such as solar power, wind power, and biofuels—hold great potential for powering society while helping humanity avoid the calamities that come with climate change. What do wind and solar power offer us? Renewables have interesting benefits and drawbacks that make their usefulness as an energy source mixed.

Even though it appears in paragraph 2, this is the essay's thesis statement. It tells what main point the essay will discuss.

Paragraph 3

Perhaps the biggest benefit of renewable energy sources is that few toxic chemicals or climate-changing substances are associated with their use. Unlike nuclear power, which produces toxic waste, and fossil fuels, which fill the atmosphere with carbon dioxide, solar and wind power produce no environmentally unfriendly by-products. Also, unlike nuclear power and oil, which are finite (meaning they come from sources that are exhaustible, or able to be permanently used up), solar and wind power are truly "renewable" in that they come from endless sources of energy—that is, it is impossible to run out of sunlight or wind. Their renewable nature

What is the topic sentence of paragraph 3? How did you recognize it?

and their lack of contribution to climate change makes them a very attractive power source. Says Sven Teske, energy expert of Greenpeace International, "Wind power will significantly reduce CO_2 emissions, which is key in the fight against dangerous climate change" (qtd. in "Wind Power Key to Fight Climate Change").

Why has the author included Sven Teske's job title?

Paragraph 4

What is the topic sentence of paragraph 4? What pieces of evidence are used to show that it is true?

Another positive aspect of renewable fuels is that they are capable of being produced in America, thereby offering the United States much-needed energy independence. Indeed, America's dependence on foreign oil not only makes the country politically vulnerable, but also the transporting of all that oil around the globe is itself a contributing factor to climate change—using oil to transport oil is like a double whammy to climate change. Solar panels, on the other hand, can be constructed at home, in the vast expanses of the American desert, or even on the rooftops of individual homes and businesses in especially sunlit states. Similarly, wind turbines can stretch the length of America's spacious plains and also be erected offshore. Corn, soy, and other matter that goes into the production of biofuels, another low-pollution energy source, can also be grown in America's vast corn belt. Each of these energy sources can help the United States develop a domestic energy economy while curbing climate change.

Paragraph 5

Renewable energy sources can also help curb climate change without taking a chunk out of Americans' wallets. Indeed, the price of wind and solar power go down every year, making them an increasingly affordable option for powering the country. In some parts of the United States, solar power costs 20 to 25 cents per kilowatt hour, while wind power—which once cost as much as 80 cents per kilowatt hour—now costs just 4 to 6 cents per kilowatt hour. In contrast, the price of oil spikes and falls erratically as a result of the dwindling supply and the difficulty of

"In contrast" is a transitional statement commonly used in compare-contrast essays. What other transitional phrases are used in this essay?

getting it from war-torn places such as the Middle East. Indeed, these are just some of the factors that caused gasoline to shoot from $2.50 a gallon in 2006 to more than $4.10 per gallon in 2008. In comparison, renewables look very affordable and stable.

Paragraph 6

But solar, wind, biofuels, and other renewable sources of energy are not without their problems. Indeed, that they do not pollute on the scale of oil does not mean they do not cause environmental damage—in fact, both power sources impact the land upon which they are erected, and sometimes gravely. In order to make use of renewable power, wind turbines or solar photovoltaic panels must be erected to harness wind or sunlight. These machines are big and would need to take up large areas of land in order to generate worthwhile quantities of power.

Paragraph 7

For example, consider that tens of thousands of acres of desert, grassland, and other habitats would need to be paved to make room for the number of panels and turbines that could capture enough light and wind to power just one city. Moreover, to sufficiently power all of the United States, it is estimated that 9 million acres of land would need to be paved with turbines and panels—an area about the size of New Hampshire and Vermont put together. Such construction would gobble up animal habitat and threaten the survival of many species while producing relatively small amounts of power. As the editors of the Colorado Springs, Colorado, newspaper the *Gazette* have asked, "How many thousands of acres of beloved 'open space' are windmill farm backers willing to obliterate along the way? . . . How much endangered species habitat consumed? . . . And how many fluffy little mountain plovers [a type of bird] are we willing to sacrifice annually to those whirling turbine blades?" In fact, the threat to wildlife from wind turbines is so high that the U.S. Fish and Wildlife Service

> Identify a piece of evidence used to support paragraph 7's main idea.

has recommended that wind turbines not be installed near wetlands, mountain ridges, or shorelines where bird populations live.

Paragraph 8

Identify the topic sentence of paragraph 8. How is it different from, but related to, the other topics discussed thus far?

Analyze this quotation. What do you think made the author want to select it for inclusion in the essay?

In addition, this damage would not even yield very much power: Renewables are notorious for not being able to deliver energy on a large enough scale required by most cities. There are also logistical concerns with erecting enough solar panels or wind turbines to provide even small amounts of power. For example, engineer Ed Hiserodt claims that a state like Arkansas would need 3,516 wind turbines to generate just 11 percent of the area's energy. But this many machines crowded together would impact wind patterns to the point where it would actually be impossible to harvest energy: "Wind turbines cannot be lined up in a row as the resulting turbulence would lower the downwind turbines to zero production if not destroy them from asymmetrical wind forces." (Hiserodt) In other words, because only so many wind turbines can be crowded together, the amount of energy this power source can generate is limited. This is the ultimate irony of renewable power—that the sheer amount of technology needed to generate useful quantities of it actually renders that power useless. As such, it cannot help mitigate the climate crisis.

Paragraph 9

The author concludes by reflecting on the pros and cons that were discussed in the essay.

Clearly, renewable sources of energy both positively and negatively contribute to the climate crisis. Given this fact, lawmakers should try to harness the benefits of renewable power while downplaying its negatives. For example, if erecting solar panels on a large scale could hurt the environment, officials should encourage their use in smaller operations such as homes and offices. Smaller-scale usage of renewable energy is likely to help curb climate change in a way that would have little or no cost to the environment without further worsening our environmental situation. Compromises such as this should take center stage in any energy policy the United States makes as it looks ahead to solving the climate crisis.

Works Cited

Hiserodt, Ed. "Blown Away." *New American* 3 Sept. 2007.

Lambrides, Mark, and Juan Cruz Monticelli. "Illuminating the Power of Renewable Energy." *Americas* May–June 2007 < http://www.oas.org/dsd/Generalassembly/Energy Final_EN.pdf > .

"Tilting at Windmills." *Colorado Springs Gazette* 25 Oct. 2003.

"Wind Power Key to Fight Climate Change." Greenpeace International. 20 Sept. 2006 < http://www.greenpeace.org/international/press/releases/wind-power-key-to-fight-climat > .

Exercise 3A: Examining Introductions and Conclusions

Every essay features introductory and concluding paragraphs that are used to frame the main ideas being presented. Along with presenting the essay's thesis statement, well-written introductions should grab the attention of the reader and make clear why the topic being explored is important. The conclusion reiterates the essay's thesis and is also the last chance for the writer to make an impression on the reader. Strong introductions and conclusions can greatly enhance an essay's effect on an audience.

The Introduction

There are several techniques that can be used to craft an introductory paragraph. An essay can start with

- an anecdote: a brief story that illustrates a point relevant to the topic;
- startling information: facts or statistics that elucidate the point of the essay;
- setting up and knocking down a position: a position or claim believed by proponents of one side of a controversy, followed by statements that challenge that claim;
- historical perspective: an example of the way things used to be that leads into a discussion of how or why things work differently now;
- summary information: general introductory information about the topic that feeds into the essay's thesis statement.

Step One

Reread the introductory paragraphs of the model essays and of the viewpoints in Section One. Identify which of the techniques described above are used in the example essays. How do they grab the attention of the reader? Are their thesis statements clearly presented?

Step Two

Write an introduction for the essay you have outlined and partially written in Exercise 2B using one of the techniques described above.

The Conclusion

The conclusion brings the essay to a close by summarizing or returning to its main ideas. Good conclusions, however, go beyond simply repeating these ideas. Strong conclusions explore a topic's broader implications and reiterate why it is important to consider. They may frame the essay by returning to an anecdote featured in the opening paragraph. Or, they may close with a quotation or refer to an event in the essay. In opinionated essays, the conclusion can reiterate which side the essay is taking or ask the reader to reconsider a previously held position on the subject.

Step Three

Reread the concluding paragraphs of the model essays and of the viewpoints in Section One. Which were most effective in driving home their arguments to the reader? What sorts of techniques did they use to do this? Did they appeal emotionally to the reader or bookend an idea or event referenced elsewhere in the essay?

Step Four

Write a conclusion for the essay you have outlined and partially written in Exercise 2B using one of the techniques described above.

Exercise 3B: Using Quotations to Enliven Your Essay

No essay is complete without quotations. Get in the habit of using quotations to support at least some of the ideas in your essays. Quotations do not need to appear in every paragraph but often enough so that the essay contains voices aside from your own. When you write, use quotations to accomplish the following:

- Provide expert advice that you are not necessarily in the position to know about.
- Cite lively or passionate passages.
- Include a particularly well-written point that gets to the heart of the matter.
- Supply statistics or facts that have been derived from someone's research.
- Deliver anecdotes that illustrate the point you are trying to make.
- Express first-person testimony.

There are a couple of important things to remember when using quotations:

- Note your sources' qualifications and biases. This way your reader can identify the person you have quoted and can put their words in a context.
- Put any quoted material within proper quotation marks. Failing to attribute quotations to their authors constitutes plagiarism, which is when an author takes someone else's words or ideas and presents them as his or her own. Plagiarism is a very serious infraction and must be avoided at all costs.

Problem One

Reread the essays presented in all sections of this book and find at least one example of each of the above quotation types.

Exercise: Write Your Own Compare-and-Contrast Five-Paragraph Essay

Using the information from this book, write your own five-paragraph compare-and-contrast essay that deals with a topic relating to climate change. You can use the resources in this book for information about issues relating to this topic and how to structure this type of essay.

The following steps are suggestions on how to get started.

Step One: Choose your topic.

The first step is to decide on what topic to write your compare-and-contrast essay. Is there any subject that particularly fascinates you? Is there an issue you strongly support or feel strongly against? Is there a topic you feel personally connected to or one that you would like to learn more about? Ask yourself such questions before selecting your essay topic. Refer to Appendix D: Sample Essay Topics if you need help selecting a topic.

Step Two: Write down questions and answers about the topic.

Before you begin writing, you will need to think carefully about what ideas your essay will contain. This is a process known as *brainstorming*. Brainstorming involves asking yourself questions and coming up with ideas to discuss in your essay. Possible questions that will help you with the brainstorming process include:

- Why is this topic important?
- Why should people be interested in this topic?
- How can I make this essay interesting to the reader?
- What question am I going to address in this paragraph or essay?
- What facts, ideas, or quotations can I use to support the answer to my question?

Questions especially for compare-and-contrast essays include:

- Have I chosen subjects that I can compare or contrast?
- What characteristics do my subjects share?
- What is different about my subjects?
- Is one subject consistently superior to another?
- Is one subject consistently inferior to another?

Step Three: Gather facts, ideas, and anecdotes related to your topic.

This book contains several places to find information, including the viewpoints and the appendices. In addition, you may want to research the books, articles, and Web sites listed in Section Three, or do additional research in your local library. You can also conduct interviews if you know someone who has a compelling story that would fit well in your essay.

Step Four: Develop a workable thesis statement.

Use what you have written down in steps two and three to help you articulate the main point or argument you want to make in your essay. It should be expressed in a clear sentence and make an arguable or supportable point.

Example:

Placing costly emissions restrictions on corporations is an unnecessary and financially devastating response to the so-called "problem" of climate change.

This could be the thesis statement of a compare-and-contrast essay that examines the benefits and drawbacks of placing emissions restrictions on corporations in an effort to curb global warming: The author will argue that because it is not certain that climate change is a bad thing, it is unwise to impose costly emissions restrictions.

Step Five: Write an outline or diagram.
  1. Write the thesis statement at the top of the outline.
  2. Write roman numerals I, II, and III on the left side of the page with A, B, and C under each numeral.
  3. Next to each roman numeral, write down the best ideas you came up with in step three. These should all directly relate to and support the thesis statement. If the essay is solely a compare or solely a contrast essay, write down three similarities or three differences between your subjects. If it is a persuasive compare-and-contrast essay, write down three reasons why one subject or argument is superior to the other.
  4. Next to each letter, write down information that supports that particular idea.

An alternative to the roman numeral outline is a diagram.

Diagrams: Alternative to Outlines

Some students might prefer to organize their ideas without using the roman numeral outline. One way to do this is to use the diagram method. Compare-and-contrast essays are especially well suited for the diagram method, which allows you to physically visualize the similarities and differences between your subjects. A possible approach would be as follows:

  a. Draw two intersecting circles in the middle of a page so that one side of each overlaps.
  b. On the left side of the page above the first circle, write "Subject A." In this circle, write all of the things that are unique to one subject.
  c. On the right side of the page above the second circle, write "Subject B." Use this circle to jot down all of the things that are unique to the other subject.
  d. In the middle of the page, above the intersection of the two circles, write "A and B." In this intersected space, write all of the things that are common to both subjects.

Step Six: Write the three supporting paragraphs.
Use your outline to write the three supporting paragraphs. Write down the main idea of each paragraph in sentence form. Do the same thing for the supporting points of information. Each sentence should support the paragraph of the topic. Be sure you have relevant and interesting details, facts, and quotations. Use transitions when you move from idea to idea to keep the text fluid and smooth. Sometimes, although not always, paragraphs can include a concluding or summary sentence that restates the paragraph's argument.

Step Seven: Write the introduction and conclusion.
See Exercise 3A for information on writing introductions and conclusions.

Step Eight: Read and rewrite.
As you read, check your essay for the following:
- ✔ Does the essay maintain a consistent tone?
- ✔ Do all paragraphs reinforce your general thesis?
- ✔ Do all paragraphs flow from one to the other? Do you need to add transition words or phrases?
- ✔ Have you quoted from reliable, authoritative, and interesting sources?
- ✔ Is there a sense of progression throughout the essay?
- ✔ Does the essay get bogged down in too much detail or irrelevant material?
- ✔ Does your introduction grab the reader's attention?
- ✔ Does your conclusion reflect on any previously discussed material or give the essay a sense of closure?
- ✔ Are there any spelling or grammatical errors?

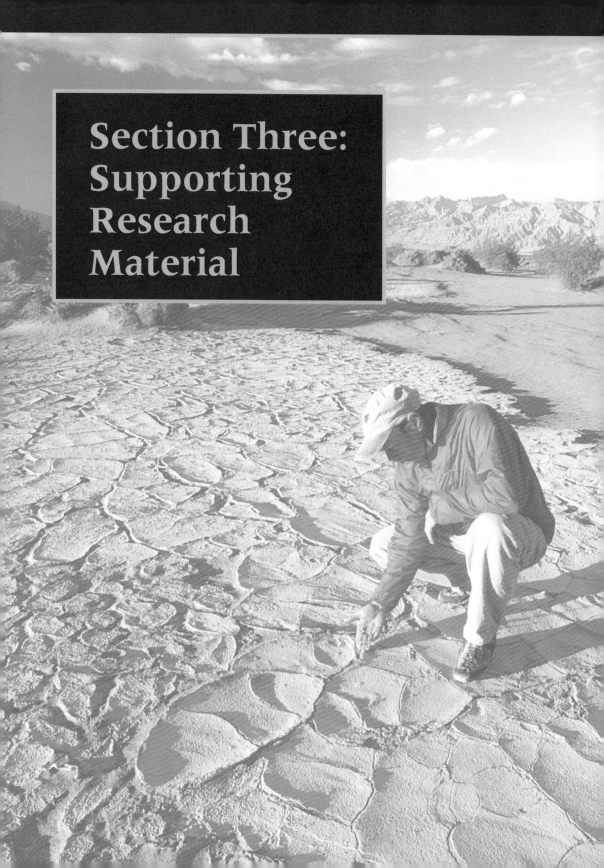

Section Three:
Supporting
Research
Material

Facts About Climate Change

Editor's Note: These facts can be used in reports to reinforce or add credibility when making important points or claims.

Facts About Climate Change

According to the Intergovernmental Panel on Climate Change (IPCC):

- During the twentieth century, the global average surface temperature increased by about 1.08°F (0.6°C).
- Global sea levels increased about 6 to 8 inches (15 to 20cm).
- The average global temperature will rise another 2° to 10°F (1.1° to 5.4°C) by 2100.
- Earth's climate will change faster in the one hundred years between 2000 and 2100 than in the last ten thousand years.
- A 2007 report by more than twenty-five hundred scientists for the IPCC concluded there is a 90 percent chance that humans are the main cause of climate change.
- The twentieth century's last two decades—the 1980s and 1990s—were the hottest in four hundred years and possibly the warmest for several millennia.
- Eleven of the past twelve years were among the warmest since 1850.
- Sea levels could rise between 7 and 23 inches (18 to 59cm) by 2100.
- Rises of just 4 inches (10cm) could flood many South Seas islands and swamp parts of Southeast Asia.
- A hundred million people live within 3 feet (1m) of sea level. Most of the global population lives in

coastal areas that could flood should sea levels rise dramatically.

- At some point in the future, rising temperatures could release additional greenhouse gases by unlocking methane currently in permafrost and undersea deposits. This would free carbon trapped in sea ice and cause increased evaporation of water and release of CO_2 into the atmosphere. This phenomenon, called the "positive feedback effect," would make global climate change become uncontrollable.

NASA's Goddard Institute for Space Studies says that average global temperatures have climbed 1.4°F (0.8°C) around the world since 1880, mostly in recent decades.

According to a multinational Arctic Climate Impact Assessment report:

- The Arctic is feeling the effects of climate change the most. Average temperatures in Alaska, western Canada, and eastern Russia have risen at twice the global average.
- Arctic ice is rapidly disappearing, and the region may have its first completely ice-free summer by 2040 or earlier.
- Glaciers and mountain snows are rapidly melting— for example, Montana's Glacier National Park now has only 27 glaciers, versus 150 in 1910.
- In the Northern Hemisphere, thaws come a week earlier in spring and freezes begin a week later.

According to *National Geographic*, coral reefs, which are very sensitive to small changes in water temperature, suffered the worst bleaching (die-off) event ever recorded in 1998. Some areas bleached at a rate of 70 percent.

Natural cycles in Earth's orbit can alter the planet's exposure to sunlight, which may explain current changes in the climate. Earth has experienced warming and cooling cycles roughly every hundred thousand years due to these orbital shifts.

According to the National Wildlife Federation, approximately 20 to 30 percent of plant and animal species are likely to be at increased risk of extinction due to global warming.

According to the *Telegraph* (London), the medieval warming period (which ended around the year 1300) was up to 5.4°F (3°C) warmer than current temperatures.

As recorded by all four agencies that track Earth's temperature, the Hadley Climate Research Unit in Britain, the NASA Goddard Institute for Space Studies in New York, the Christy group at the University of Alabama, and Remote Sensing Systems Inc. in California:

- The global average temperature cooled by about 1.3°F (0.7°C) in 2007.
- This is the fastest temperature change in the instrumental record and equaled the global average temperature of 1930.

According to the organization The Climate Trust:

- 1998 was the hottest year on record, followed by 2002, 2003, and 2004.
- The ten hottest years ever documented have all occurred since 1990.
- The average global surface temperature in the year 2100 will likely be 2.5° to 10.8°F (1.4° to 6°C) higher than in 1990.
- The average Arctic winter temperature has increased by 11°F (6°C).
- Coastal glaciers in Greenland are undergoing rapid thinning by as much as 3 feet (1m) per year.

- Glaciers in Glacier National Park are receding so rapidly that the park is expected to have no glaciers within several decades.

Facts About Greenhouse Gases

Greenhouse gases help keep the planet warm—but in excess, they make it get too warm. Some greenhouse gases occur naturally, and others are produced from human activity.

Carbon dioxide or CO_2 is the most significant greenhouse gas associated with climate change. It is released into the air as a result of human activities, such as burning fossil fuels. Carbon dioxide is the main contributor to climate change.

As of 2008, the world's top three CO_2 emitters were:
1) China
2) The United States
3) India

Methane is produced when vegetation is burned, digested, or rotted with no oxygen present. Garbage dumps, rice paddies, and grazing cows and other livestock release large amounts of methane.

Nitrous oxide is found naturally in the environment. Human activities, however, are increasing the amounts found in the atmosphere. Nitrous oxide is put into the air when chemical fertilizers and manure are used.

Halocarbons are a family of chemicals that include CFCs, which also damage the ozone layer. Human-made chemicals that contain chlorine and fluorine also release halocarbons.

According to *National Geographic*, humans are releasing carbon dioxide into the atmosphere faster than plants and oceans can absorb it. These gases remain in the atmosphere for years, which means that even if such emissions were eliminated today, it would not immediately have any positive effect on climate change.

Facts About the Kyoto Treaty

- The Kyoto Treaty or Protocol is an international framework for reducing greenhouse gas emissions to prevent man-made climate change. It makes signatories legally obligated to curtail their greenhouse gas emissions.
- It was first negotiated in Kyoto, Japan, in December 1997.
- The treaty has been ratified by 181 countries as of May 2008.
- The United States is the only industrialized nation that has not ratified the treaty. The United States has refused to do so because it claims the country would be hurt economically.
- As one of the world's largest CO_2 emitters, the United States has been heavily criticized for failing to ratify the Kyoto Treaty.
- The organization The Climate Trust says that Kyoto will halt rising world temperatures by only 0.2°F (0.1°C).

Opinions on Climate Change

According to a 2008 poll by Ipsos MORI:

- 60 percent of British people think that scientific experts are not sure whether climate change is human-caused.
- 40 percent think that climate change might not be as bad as people say.
- People who are most worried about climate change are more likely to have a college degree and have a higher income than the general population.

According to a 2007 poll by the British Broadcasting Company:

- 79 percent of people in twenty-one nations believe human activity causes global warming.
- 90 percent say action is needed to address global warming.
- 65 percent believe immediate action is needed.

According to a 2008 Harris poll:
- Nearly 66 percent of Americans want the next president to initiate strong action on climate change.
- 40 percent think that global warming could threaten national security if it continues unchecked.

A 2008 Gallup poll found that:
- 21 percent say they understand the issue of global warming "very well."
- 59 percent say they understand it "fairly well."
- 20 percent say they understand it "not very well" or not at all.
- 61 percent of Americans say the effects of global warming have already begun.
- 14 percent say it will begin within their lifetime.
- 13 percent say it will begin not in their lifetime but will affect future generations.
- 11 percent say it will never happen.
- 40 percent say global warming poses a threat to their way of life.
- 58 percent say global warming does not pose a threat to their way of life.
- 37 percent say they worry about global warming a great deal.
- 29 percent say they worry about global warming a fair amount.
- 33 percent say they do not worry about global warming at all.
- 34 say that immediate drastic action is needed to combat global warming.
- 52 say that some additional action is needed.
- 13 say that no additional action is needed other than what is currently being undertaken.

PROPERTY OF
MONTCALM
COMMUNITY COLLEGE

Finding and Using Sources of Information

No matter what type of essay you are writing, it is necessary to find information to support your point of view. You can use sources such as books, magazine articles, newspaper articles, and online articles.

Using Books and Articles

You can find books and articles in a library by using the library's computer or cataloging system. If you are not sure how to use these resources, ask a librarian to help you. You can also use a computer to find many magazine articles and other articles written specifically for the Internet.

You are likely to find a lot more information than you can possibly use in your essay, so your first task is to narrow it down to what is likely to be most usable. Look at book and article titles. Look at book chapter titles, and examine the book's index to see if it contains information on the specific topic you want to write about. (For example, if you want to write about how biofuels could impact climate change and you find a book about alternative energy sources, check the chapter titles and index to be sure it contains information about biofuels before you bother to check out the book.)

For a five-paragraph essay, you do not need a great deal of supporting information, so quickly try to narrow down your materials to a few good books and magazines or Internet articles. You do not need dozens. You might even find that one or two good books or articles contain all the information you need.

You probably do not have time to read an entire book, so find the chapters or sections that relate to your topic, and skim these. When you find useful information, copy it onto a note card or notebook. You should look for supporting facts, statistics, quotations, and examples.

Using the Internet

When you select your supporting information, it is important that you evaluate its source. This is especially important with information you find on the Internet. Because nearly anyone can put information on the Internet, there is as much bad information as good information. Before using Internet information—or any information—try to determine if the source seems to be reliable. Is the author or Internet site sponsored by a legitimate organization? Is it from a government source? Does the author have any special knowledge or training relating to the topic you are looking up? Does the article give any indication of where its information comes from?

Using Your Supporting Information

When you use supporting information from a book, article, interview, or other source, there are three important things to remember:

1. *Make it clear whether you are using a direct quotation or a paraphrase.* If you copy information directly from your source, you are quoting it. You must put quotation marks around the information and tell where the information comes from. If you put the information in your own words, you are paraphrasing it.

Here is an example of a using a quotation:

Reporter Gregg Easterbrook suggests that climate change need not be all bad for the planet—in fact, many nations could even benefit from it. "If the global climate continues changing, many people and nations will find themselves in possession of land and resources of rising value," he explains. "Climate change increases the supply of land by warming currently frosty areas while throwing the amount of desirable land into tremendous flux." (53)

Here is an example of a brief paraphrase of the same passage:

Reporter Gregg Easterbrook suggests that climate change need not be all bad for the planet—in fact, many nations could even benefit from it. As the climate changes, currently desirable places could become unbearably hot, while previously frigid places could warm up enough to be temperate and pleasant. In this way, the locations of valuable, inhabitable land could be radically altering the ways of living for populations all over the planet.

2. *Use the information fairly.* Be careful to use supporting information in the way the author intended it. For example, it is unfair to quote an author as saying, "Renewable resources such as solar and wind power are great sources of energy," when he or she intended to say, "Renewable resources such as solar and wind power are great sources of energy—provided the sun is out and the wind is blowing." This is called taking information out of context. This is using supporting evidence unfairly.

3. *Give credit where credit is due.* Giving credit is known as citing. You must use citations when you use someone else's information, but not every piece of supporting information needs a citation.

   - If the supporting information is general knowledge—that is, it can be found in many sources—you do not have to cite your source.
   - If you directly quote a source, you must cite it.
   - If you paraphrase information from a specific source, you must cite it.

If you do not use citations where you should, you are *plagiarizing*—or stealing—someone else's work.

Citing Your Sources

There are a number of ways to cite your sources. Your teacher will probably want you to do it in one of three ways:
   - Informal: As in the examples in number 1 above, tell where you got the information as you present it in the text of your essay.

- Informal list: At the end of your essay, place an unnumbered list of all the sources you used. This tells the reader where, in general, your information came from.
- Formal: Use numbered footnotes or endnotes. Footnotes or endnotes are generally placed at the end of an article or essay, although they may be placed elsewhere depending on your teacher's requirements.

Works Cited

Easterbrook, Gregg. "Global Warming: Who Loses—and Who Wins?" *Atlantic Monthly* April 2006: 52–61.

Climate Change Is Man-Made

Juliet Eilperin

In the following essay Juliet Eilperin reports on a 2006 study by the Intergovernmental Panel on Climate Change (IPCC), which concluded that global warming is a man-made phenomenon. The report, which was written by hundreds of scientists from 113 countries, concluded with 90 percent certainty that human industry is responsible for the global rise in temperatures that has occurred over the last fifty years. Scientists predicted that the world is due for a severe rise in temperature that will cause species to become extinct and sea levels to rise dangerously. Eilperin reports that officials are mixed on how to proceed—some want drastic cuts made to greenhouse gas emissions, while others believe it is reckless to make such changes without being sure of their economic and environmental consequences. Eilperin concludes that from the IPCC's perspective, climate change is a human-caused phenomenon that is not going away anytime soon.

Eilperin is a staff writer for the *Washington Post*.

Consider the following questions:

1. What would a temperature increase of more than 3.6°F (2°C) do to the planet, as reported by Eilperin?
2. How did the 2006 Intergovernmental Panel on Climate Change report on global warming differ from the organization's 2001 report, in regard to whether global warming is human-caused?
3. Who is Gerald Meehl, and what was his take on the 2006 IPCC report on climate change?

Juliet Eilperin, "Humans Faulted for Global Warming," washingtonpost.com, February 3, 2007. Reproduced by permission.

Using MLA Style to Create a Works Cited List

You will probably need to create a list of works cited for your paper. These include materials that you quoted from, relied heavily on, or consulted to write your paper. There are several different ways to structure these references. The following examples are based on Modern Language Association (MLA) style, one of the major citation styles used by writers.

Book Entries

For most book entries you will need the author's name, the book's title, where it was published, what company published it, and the year it was published. This information is usually found on the inside of the book. Variations on book entries include the following:

A book by a single author:
> Friedman, Thomas. *Hot, Flat, and Crowded: Why We Need a Green Revolution—and How It Can Renew America.* New York: Farrar, Straus and Giroux, 2008.

Two or more books by the same author:
> Pollen, Michael. *Botany of Desire: A Plant's-Eye View of the World.* New York: Random House, 2002.
> ———. *The Omnivore's Dilemma: A Natural History of Four Meals.* New York: Penguin Books, 2006.

A book by two or more authors:
> Esposito, John L., and Dalia Mogahed. *Who Speaks for Islam? What a Billion Muslims Really Think.* Washington, DC: Gallup Press, 2008.

A book with an editor:
> Skancke, Jennifer S., ed. *Introducing Issues with Opposing Viewpoints: Stem Cell Research.* Detroit: Greenhaven, 2008.

Periodical and Newspaper Entries

Entries for sources found in periodicals and newspapers are cited a bit differently from books. For one, these sources usually have a title and a publication name. They also may have specific dates and page numbers. Unlike book entries, you do not need to list where newspapers or periodicals are published or what company publishes them.

An article from a periodical:
> Aldhous, Peter. "China's Burning Ambition." *Nature* 30 June 2005: 1152–55.

An unsigned article from a periodical:
> "Contraception in Middle School?" *Harvard Crimson* 21 Oct. 2007.

An article from a newspaper:
> Cunningham, Roseanna. "Care, not Euthanasia, Is the Answer to the 'Problem' of the Elderly." *Sunday Times* (London) 20 July 2008: 21.

Internet Sources

To document a source you found online, try to provide as much information on it as possible, including the author's name, the title of the document, date of publication or of last revision, the URL, and your date of access.

A Web source:
> Mieszkowski, Katharine. "Plastic Bags Are Killing Us." Salon.com 10 Aug. 2007. September 9, 2008 http://www.salon.com/news/feature/2007/08/10/plastic_bags/index.html.

Your teacher will tell you exactly how information should be cited in your essay. Generally, the very least information needed is the original author's name and the name of the article or other publication.

Be sure you know exactly what information your teacher requires before you start looking for your supporting information so that you know what information to include with your notes.

Sample Essay Topics

Climate Change Is Occurring
Climate Change Is Not Occurring
Humans Are Causing Climate Change
Humans Are Not Causing Climate Change
Climate Change Is Bad for the Planet
Climate Change Is Not Bad for the Planet
Climate Change Threatens Crop Growth
Climate Change Encourages Crop Growth
Climate Change Will Cause World Hunger
Climate Change Will Help Feed the World
Climate Change Threatens Biodiversity
Climate Change Encourages Biodiversity
Climate Change Will Cause Sea Levels to Rise
Climate Change Will Not Cause Sea Levels to Rise
Climate Change Will Increase Disease
Climate Change Will Decrease Disease
Climate Change Is Causing Devastating Hurricanes
Climate Change Is Not Causing Devastating Hurricanes
A Long History of Climate Change and Its Effects

Topics for Compare/Contrast Essays

Examining the Causes of Climate Change
Examining Solutions to Climate Change
Fossil Fuel Use Is the Biggest Contributor to Climate Change
Fossil Fuel Use Is Not Likely Contributing to Climate Change
Nuclear Power Can Curb Climate Change Better than
    Renewable Energy
Renewable Energy Can Curb Climate Change Better than
    Nuclear Power
Switching to a Local Food Economy Is the Best Way to
    Curb Climate Change
Comparing Methods of Determining Climate Change

Organizations to Contact

The editors have compiled the following list of organizations concerned with the issues debated in this book. The descriptions are derived from materials provided by the organizations. All have publications or information available for interested readers. The list was compiled on the date of publication of the present volume; the information provided here may change. Be aware that many organizations take several weeks or longer to respond to queries, so allow as much time as possible.

Adelphi Research
Caspar-Theyss-Str. 14a, 14193 Berlin • e-mail: office@ adelphi-research.de • Web site: www.adelphi-research.de/

Adelphi Research is an independent, nonprofit institute that develops and implements innovative sustainable development strategies. Its mission is to increase awareness and understanding of the political, economic, and technological forces driving global change. It publishes numerous papers on climate change and security.

Climate Solutions
219 Legion Way, Ste. 201, Olympia, WA 98501-1113 • (360) 352-1763 • e-mail: info@climatesolutions.org • Web site: http://climatesolutions.org

Climate Solutions' mission is to stop global warming at the earliest possible point by helping the Northwest region of the United States develop practical and profitable solutions.

Competitive Enterprise Institute (CEI)
1001 Connecticut Ave. N W, Ste. 1250, Washington, DC 20036 (202) 331-1010 • fax: (202) 331-0640 • e-mail: info@cei.org Web site: www.cei.org

CEI is a nonprofit organization dedicated to the principles of free enterprise and limited government. Rather than promoting government regulation, it advocates removing governmental barriers and using private incentives to protect the environment.

Cooler Heads Coalition
c/o Consumer Alert 1001 Connecticut Ave. NW, Ste. 1128, Washington, DC 22036 • (202) 467-5809 • fax: (202) 467-5814

The Cooler Heads Coalition is a subgroup of the National Consumer Coalition and was founded by that group to debunk what they regard as myths of global warming. Features of the Web site include economic arguments against the Kyoto Protocol and other climate change policy documents as well as regular legislative updates.

Friends of the Earth
1717 Massachusetts Ave. NW, Washington, DC 20036 (877) 843-8687 • fax: (202) 783-0444 • e-mail: foe@foe .org • Web site: www.foe.org

Friends of the Earth is an activist organization dedicated to protecting the planet from environmental disaster. They have worked for renewable energy development, limitation of carbon dioxide emissions, and protection of the environment.

The George C. Marshall Institute
1625 K St. NW, Washington, DC 20006 • (202) 296-9655 fax: (202) 296-9714 • e-mail: info@marshall.org • Web site: www.marshall.org

The institute is a research group that provides scientific and technical advice and promotes scientific literacy on matters that have an impact on public policy. The institute's publications include the book *Shattered Consensus: The True State of Global Warming* and many studies, including "Natural Climate Variability" and "Climate Issues and Questions."

Global Warming International Center (GWIC)

PO Box 50303, Palo Alto, CA 94303 • (630) 910-1551 • fax: (630) 910-1561 • Web site: www.globalwarming.net

GWIC is an international body that disseminates information on science and policy concerning global warming. It serves both governmental and nongovernmental organizations as well as industries in more than one hundred countries. The center sponsors unbiased research supporting the understanding of global warming.

Greenpeace USA

702 H St. NW, Ste. 300, Washington, DC 20001 • (202) 462-1177 • Web site: www.greenpeace.org/usa

Greenpeace has long campaigned against environmental degradation, urging government and industry to stop climate change, protect forests, save the oceans, and stop the nuclear threat. It uses controversial direct-action techniques and strives for media coverage of its actions in an effort to educate the public. Greenpeace publishes reports and pamphlets about climate and energy, forests, and genetic engineering of crops.

The Heartland Institute

19 S. LaSalle St., Ste. 903, Chicago, IL 60603 • (312) 377-4000 • Web site: www.heartland.org

The Heartland Institute is a national nonprofit research and education organization. Its stated mission is to discover and promote free-market solutions to social and economic problems, including market-based approaches to environmental protection.

The Intergovernmental Panel on Climate Change (IPCC)

c/o World Meteorological Organization, 7bis Ave. de la Paix C.P. 2300, Geneva 2, Switzerland CH-1211 • +41-22-730-8208 • e-mail: ipcc-sec@wmo.int • Web site: www.ipcc.ch

Established in 1988, the IPCC's role is to assess the scientific, social, and economic information relevant for the understanding of the risk of human-induced climate change.

Met Office Hadley Centre

Met Office, FitzRoy Rd., Exeter, Devon EX1 3PB, United Kingdom • e-mail: enquiries@metoffice.gov.uk • Web site: www.metoffice.gov.uk

The Met Office Hadley Centre is the United Kingdom's official center for climate change research, focusing particularly on scientific issues. This agency advises the British government on how to respond to climate change issues.

The Pembina Institute for Appropriate Development

Box 7558, Drayton Valley, AB T7A 1S7, Canada • (780) 542-6272 • fax: (780) 542-6464 • Web site: www.pembina.org

The Pembina Institute is an independent, not-for-profit environmental policy research and education organization. Its major policy research and education programs are in the areas of sustainable energy, climate change, environmental governance, ecological fiscal reform, sustainability indicators, and the environmental impacts of the energy industry. The institute pioneers practical solutions to issues affecting human health and the environment across Canada. It publishes numerous papers and reports relating to climate change issues.

Rainforest Action Network (RAN)

221 Pine St., 5th Floor, San Francisco, CA 94104 • (415) 398-4404 • e-mail: answers@ran.org • Web site: www.ran.org

RAN works to preserve the world's rain forests through activism addressing the logging and importation of tropical timber, cattle ranching in rain forests, and the rights of indigenous rain forest peoples. It also seeks to educate the public about the environmental effects of tropical hardwood logging. RAN's publications include the monthly bulletin *Action Report* and the semiannual *World Rainforest Report*.

Reason Foundation

3415 S. Sepulveda Blvd., Ste. 400, Los Angeles, CA 90034-6064 • (310) 391-2245 • Web site: www.reason.org

The Reason Foundation is a national public policy research organization. It specializes in a variety of policy areas, including the environment, education, and privatization. The foundation publishes the monthly magazine *Reason* and the books *Global Warming: The Greenhouse, White House, and Poorhouse Effect*; *The Case Against Electric Vehicle Mandates in California*; and *Solid Waste Recycling Costs—Issues and Answers*.

Sierra Club

85 Second St., San Francisco, CA 94105 • (415) 977-5500 fax: (415) 977-5799 • e-mail: info@sierraclub.org • Web site: www.sierraclub.org

The Sierra Club is a grassroots organization that promotes the protection and conservation of natural resources. It publishes the bimonthly magazine *Sierra* and e-mail newsletter the *Insider*, as well as numerous books and pamphlets.

Union of Concerned Scientists (UCS)

2 Brattle Sq., Cambridge, MA 02238-9105 • (617) 547-5552 Web site: www.ucsusa.org

UCS works to advance responsible public policy in areas where science and technology play a vital role. Its programs focus on safe and renewable energy technologies, transportation reform, arms control, and sustainable agriculture. UCS publications include the quarterly magazine *Nucleus*, the briefing papers "Motor-Vehicle Fuel Efficiency and Global Warming and Global Environmental Problems: A Status Report," and the book *Cool Energy: The Renewable Solution to Global Warming*.

United Nation Environment Programme (UNEP)

United Nations Ave., Gigiri, PO Box 30552, 00100 Nairobi, Kenya • +254-2-7621234 • e-mail: unepinfo@unep.org Web site: www.unep.org

The mission of the UNEP is to provide leadership and encourage partnership in caring for the environment by inspiring, informing, and enabling nations and peoples to improve their quality of life without compromising that of future generations.

World Resources Institute (WRI)

10 G St. NE, Ste. 800, Washington, DC 20002 • (202) 729-7600 • fax: (202) 729-7610 • e-mail: rspeight@wri.org Web site: www.wri.org

WRI conducts policy research on global resources and environmental conditions. It publishes books, reports, and papers; holds briefings, seminars, and conferences; and provides the print and broadcast media with new perspectives and background materials on environmental issues. The institute's books include *Climate Science 2005: Major New Discoveries*.

Worldwatch Institute

1776 Massachusetts Ave. NW, Washington, DC 20036 (202) 452-1999 • fax: (202) 296-7365 • Web site: www .worldwatch.org

Worldwatch is a research organization that analyzes and focuses attention on global problems, including environmental concerns such as global warming and the relationship between trade and the environment. It compiles the annual *State of the World* anthology and publishes the bimonthly magazine *World Watch* and the Worldwatch Paper Series, which includes "Mainstreaming Renewable Energy in the 21st Century."

Bibliography

Books

Campbell, Kurt M., *Climatic Cataclysm: The Foreign Policy and National Security Implications of Climate Change*. Washington, DC: Brookings Institution Press, 2008.

Dessler, Andrew E., and Edward A. Parson, *The Science and Politics of Global Climate Change: A Guide to the Debate*. New York: Cambridge University Press, 2006.

DiMento, Joseph F.C., and Pamela Doughman, eds., *Climate Change: What It Means for Us, Our Children, and Our Grandchildren*. Cambridge, MA: MIT Press, 2007.

Emanuel, Kerry, *What We Know About Climate Change*. Cambridge, MA: MIT Press, 2007.

Flannery, Tim, *The Weather Makers: How Man Is Changing the Climate and What It Means for Life on Earth*. New York: Atlantic Monthly Press, 2005.

Gore, Al, *An Inconvenient Truth*. New York: Rodale, 2006.

Kolbert, Elizabeth, *Field Notes from a Catastrophe*. New York: Bloomsbury USA, 2006.

Linden, Eugene, *The Winds of Change: Climate, Weather, and the Destruction of Civilizations*. New York: Simon & Schuster, 2007.

Spencer, Roy, *Climate Confusion: How Global Warming Hysteria Leads to Bad Science, Pandering Politicians and Misguided Policies That Hurt the Poor*. New York: Encounter, 2008.

Sweet, William, *Kicking the Carbon Habit: Global Warming and the Case for Renewable and Nuclear Energy*. New York: Columbia University Press, 2006.

Periodicals

Avery, Dennis T., and H. Sterling Burnett, "Global Warming Famine—or Feast?" National Center for Policy Analysis,

Brief Analysis no. 517, May 19, 2005. www.ncpa.org/pub /ba/ba517/.

Ball, Timothy, "Global Warming: The Cold, Hard Facts?" *Canada Free Press*, February 5, 2007. www.canadafree press.com/2007/global-warming020507.htm.

Bloom, Noah S., "It's Getting Hot in Here: Global Warming Hits the Hub," *Harvard Crimson*, June 7, 2006.

Brown, O., A. Hammill, and R. McLeman, "Climate Change as the 'New' Security Threat: Implications for Africa," *International Affairs*, vol. 83, no. 6, December 10, 2007. www.iisd.org/pdf/2007/climate_security_threat_africa .pdf.

Carter, Robert M., "Human-Caused Global Warming," Center for Science and Public Policy, February 2007 www.ff.org/centers/csspp/pdf20070330_carter.pdf.

Center for Strategic and International Studies and Center for a New American Security, "The Age of Consequences: The Foreign Policy and National Security Implications of Global Climate Change," November 2007. www.csis .org/media/csis/pubs/071105_ageofconsequences.pdf.

Center for Strategic and International Studies and Center for a New American Security, "The Myth of Dangerous Human-Caused Climate Change," AusIMM New Leaders' Conference, May 2–3, 2007. http://members. iinet.au/~glrmc/2007%2005-03%20AusIMM%20cor- rected.pdf.

De Weese, Tom, "Global Warming: The Other Side of the Story," *Capitalism Magazine*, May 19, 2006.

Doyle, Alister, "Global Warming: Could Spur 21st Century Conflicts," Common Dreams.org, April 16, 2007. www .commondreams.org/archive/2007/04/16/552/.

Dupont, Alan, and Graeme Pearman, "Heating Up the Planet: Climate Change and Security," Lowly Institute for International Policy, Paper 12, 2006. http://lowly institute.richmedia-server.com/docs/AD_GP_Climate Change.pdf.

Dye, Lee, "Global Climate Change Is Happening Now: Scientists Fear Global Warming Is Irreversible and Its Effects Possibly Disastrous," ABC News, July 12, 2006. http://abcnews.go.com/Technology/Story?id = 2182824&page = l.

Environmental Defense Fund, "Global Warming Myths and Facts," 2008. www.edf.org/page.cfm?tagID = 1011.

Evans, Gareth, "Conflict Potential in a World of Climate Change," International Crisis Group, August 29, 2008. www.crisisgroup.org/home/index.cfm?id = 5648&l = l.

Financial Times, "Climate Change Is Not a Global Crisis—That Is the Problem," April 17, 2007. www.ft.com/cms/s/0/5eb8f012-ec80-11db-a112e-000b5df10621.html.

Grossman, Elizabeth, "The Big Melt," *Earth Island Journal*, Summer 2008.

Hertsgaard, Mark, "Nuclear Energy Can't Solve Global Warming," *San Francisco Chronicle*, August 7, 2005.

Intergovernmental Panel on Climate Change, Climate Change 2007, Summary for Policymakers, 2007. www.ipcc.ch/pdf/assessment-report/syr/ar4_syr_spm.pdf.

Johnson, Neil, "Hurricane Predictions Off Track as Tranquil Season Wafts Away," *Tampa Tribune*, November 27, 2006. www.tbo.com/news/metro/MGBHKNBEOVE.html.

Jowit, Juliette, "Poll: Most Britons Doubt Cause of Climate Change," *Guardian* (Manchester, UK), June 22, 2008. www.guardian.co.uk/environment/2008/jun/22/climatechange.carbonemissions.

Landrith, George, "Climate Change Hysteria and Al Gore's 'Chicken Little' Scare Tactics," Frontiers of Freedom, September 14, 2007. www.ff.org/index.php?option = com_content&task = view&id = 373&Itemid = 1.

Lawson, Nigel, "The Economics and Politics of Climate Change: An Appeal to Reason," Centre for Policy Studies, November 1, 2006.

Lomborg, Bjorn, "Stern Review: The Dodgy Numbers Behind the Latest Warming Scare," *Wall Street Journal*, November 2, 2006. www.opinionjournal.com/extra/?id = 110009182.

Micallef, Charles, and B. Pharm, "Global Warming—Will Malta Vanish from the Face of the Earth?" *Times of Malta*, May 21, 2006.

Monckton, Christopher, "Wrong Problem, Wrong Solution," *Telegraph* (London), November 13, 2006. www.telegraph.co.uk/news/uknews/1533912/Wrong-problem%2C-wrong-solution.html.

Murdock, Deroy, "Chill Out on Climate Hysteria: The Earth Is Currently Cooling," *National Review*, May 2, 2008. http://article.nationalreview.com/?q = OTc1Mz djOWEwMWUyNGMwYzkxMjMzZWIzMjE5NDc3MG Q = &w = MA = = .

Reid, John, "Water Wars: Climate Change May Spark Conflict," *Independent* (UK), February 28, 2006. www.independent.co.uk/environment/water-wars-climate-change-may-spark-conflict-467957.html.

Schmitt, Jerome J., "Numerical Models, Integrated Circuits and Global Warming Theory," *American Thinker*, February 28, 2007. www.americanthinker.com/2007/02/numerical_models_integrated_ci.html.

Singer, S. Fred, "Global Warming: Man-Made or Natural?" Center for Economic Freedom, September 2007. www.texaspolicy.com/pdf/2007-09-PP24-globalwarming-singer.pdf.

Stern, Nicholas, "The Stern Review on the Economics of Climate Change," HM Treasury, October 2006. http://www.hm-treasury.gov.uk/independent_reviews/stern_review_economics_climate_change/stern_review_report.cfm.

Steyn, Mark, "Climate Change Myth," *Australia*, September 11, 2006. www.freerepublic.com/focus/f-news/1555 298/posts.

Suprynowicz, Vin, "Global Warming," Lew Rockwell.com, February 28, 2007. www.lewrockwell.com/suprynowicz/suprynowicz60.html2007.

Wall Street Journal, "Climate of Opinion: The Latest U.N. Report Shows the 'Warming' Debate Is Far from Settled," February 7, 2007. www.opinionjournal.com/editorial/feature.html?id = 110009625.

Web Sites

Environment, Conflict, and Cooperation Platform (www.ecc-platform.org). This site functions as a clearinghouse and platform of exchange on issues relating to the environment, conflict, and cooperation. It offers papers and other information on the climate crisis, weather, water, energy, and other environmental and security issues.

Friends of Science (www.friendsofscience.org). This Web site contains numerous links to articles debunking theories of man-made climate change. The organization's position is that global warming is a natural phenomenon.

German Advisory Council on Global Change (www.wbgu.de/wbgu_home_engl.html). The job of this council is to analyze and report on global environment and development problems. Its publications on climate change are widely read and respected.

Institute for Environmental Security (www.envirosecurity.org). This site features the *EnviroSecurity Action Guide*, a database of organizations, initiatives, publications, and Web resources related to climate change, environment, security, and sustainable development.

Real Climate (www.realclimate.org). This Web site, run by climatologists, provides up-to-date commentary on the latest news about climate science.

U.S. Environmental Protection Agency's Page on Climate Change (www.epa.gov/climatechange). This page, run by the U.S. government, explains the official U.S. position on climate change and its resulting policies. It offers comprehensive information on the issue of climate change in a way that is accessible and meaningful to communities, individuals, businesses, states and localities, and governments.

World Climate Report (www.worldclimatereport.com). This climate change blog argues that global warming is not to be feared. The site includes opinion pieces critical of the science behind global warming and its alleged impacts.

Index

Picture Credits

Maury Aaseng, 16-17, 36-37, 44, 48, 53, 61

© age fotosock/SuperStock, 51

AP Images, 35, 39, 42, 55

© Ulrich Doering/Alamy, 28

© Suzanne Long/Alamy, 10

© John MacPherson/Alamy, 27

© Tim McGuire/Corbis, 20

Finbarr O'Reilly/Reuters/Landov, 62

Photo by Ami Vitale/Getty Images, 59

© Wolfgang Polzer/Alamy, 45

© Norman Price/Alamy, 9

© SuperStock, Inc./SuperStock, 15

© 2007 by Eric Allie and PoliticalCartoons.com, 30

About the Editor

Lauri S. Friedman earned her bachelor's degree in religion and political science from Vassar College in Poughkeepsie, New York. Her studies there focused on political Islam. Friedman has worked as a nonfiction writer, a newspaper journalist, and an editor for more than eight years. She has extensive experience in both academic and professional writing settings.

Lauri is the founder of LSF Editorial, a writing and editing business in San Diego. Her clients include Greenhaven Press, for whom she has edited and authored numerous publications on controversial social issues such as oil, the Internet, the Middle East, democracy, pandemics, and obesity. Every book in the *Writing the Critical Essay* series has been under her direction or editorship, and she has personally written more than eighteen titles in the series. She was instrumental in the creation of the series and played a critical role in its conception and development.